Biorefinery Production of Fuels and Platform Chemicals

Scrivener Publishing
100 Cummings Center, Suite 541J
Beverly, MA 01915-6106

Publishers at Scrivener
Martin Scrivener (martin@scrivenerpublishing.com)
Phillip Carmical (pcarmical@scrivenerpublishing.com)

Biorefinery Production of Fuels and Platform Chemicals

Edited by
Prakash Kumar Sarangi

WILEY

This edition first published 2023 by John Wiley & Sons, Inc., 111 River Street, Hoboken, NJ 07030, USA and Scrivener Publishing LLC, 100 Cummings Center, Suite 541J, Beverly, MA 01915, USA

For more information about Scrivener publications please visit www.scrivenerpublishing.com.

Wiley Global Headquarters
111 River Street, Hoboken, NJ 07030, USA

For details of our global editorial offices, customer services, and more information about Wiley products visit us at www.wiley.com.

Library of Congress Cataloging-in-Publication Data

ISBN 9781119724728

Cover image: Ethanol Plant: Gloria P. Meyerle | Dreamstime.com,
Flower Field: ID 181127767 © Jochenschneider | Dreamstime.com
CO2: Pop Nukoonrat | Dreamstime.com, Renewable Resources: Mario Kelichhaus | Dreamstime.com
Cover design by Kris Hackerott

Set in size of 11pt and Minion Pro by Manila Typesetting Company, Makati, Philippines

Printed in the USA

10 9 8 7 6 5 4 3 2 1

Dedication

Dedicated
to
My beloved Lord

Contents

List of Contributors

Abbhijith H
Department of Mechanical Engineering, Amrita School of Engineering, Coimbatore, Amrita Vishwa Vidyapeetham, India

Anushri Keshri
Institute of Biotechnology, Amity University Chhattisgarh, Raipur, Chhattisgarh, India

Balasubramanian Velramar
Institute of Biotechnology, Amity University Chhattisgarh, Raipur, Chhattisgarh, India

Harshini G V
Department of Mechanical Engineering, Amrita School of Engineering, Coimbatore, Amrita Vishwa Vidyapeetham, India

Harit Jha
Department of Biotechnology, Guru Ghasidas Vishwavidyalaya, Bilaspur, Chhattisgarh, India

Krushna Prasad Shadangi
Department of Chemical Engineering, Veer Surendra Sai University of Technology, Burla, Odisha, India

Meenakshi Jha
Institute of Biotechnology, Amity University Chhattisgarh, Raipur, Chhattisgarh, India

Musunuri Shanmukha Vardhan
Department of Mechanical Engineering, Amrita School of Engineering, Coimbatore, Amrita Vishwa Vidyapeetham, India

Neha Bothra
Department of Biotechnology, GIT, Gandhi Institute of Technology and Management (GITAM) Deemed to be University, Rushikonda, Visakhapatnam, India

Neha Namdeo
Department of Biotechnology, Guru Ghasidas Vishwavidyalaya, Bilaspur, Chhattisgarh, India

P. Maniharika
Department of Biotechnology, GIT, Gandhi Institute of Technology and Management (GITAM) Deemed to be University, Rushikonda, Visakhapatnam, India

Pamidimarri D. V. N. Sudheer
Institute of Biotechnology, Amity University Chhattisgarh, Raipur, Chhattisgarh, India

Prakash Kumar Sarangi
College of Agriculture, Central Agricultural University, Imphal, Manipur, India

Rajesh K. Srivastava
Department of Biotechnology, GIT, Gandhi Institute of Technology and Management (GITAM) Deemed to be University, Rushikonda, Visakhapatnam, India

Shankar Swarup Das
Department of Farm Machinery and Power Engineering, College of Agricultural Engineering and Post-Harvest Technology (Central Agricultural University), Ranpool, India

Sakthivel R
Department of Mechanical Engineering, Amrita School of Engineering, Coimbatore, Amrita Vishwa Vidyapeetham, India

Shalini Pandey
Institute of Biotechnology, Amity University Chhattisgarh, Raipur, Chhattisgarh, India

Shivam Pandey
Institute of Biotechnology, Amity University Chhattisgarh, Raipur, Chhattisgarh, India

Sneha Kumari
Institute of Biotechnology, Amity University Chhattisgarh, Raipur, Chhattisgarh, India

Sushma Chauhan
Institute of Biotechnology, Amity University Chhattisgarh, Raipur, Chhattisgarh, India

Tanushree Baldeo Madavi
Institute of Biotechnology, Amity University Chhattisgarh, Raipur, Chhattisgarh, India

Trinath Biswal
Department of Chemistry, Veer Surendra Sai University of Technology, Burla, Odisha, India

Vargobi Mukherjee
Institute of Biotechnology, Amity University Chhattisgarh, Raipur, Chhattisgarh, India

Preface

Economic growth through extensive use of fossil resources is generally considered as unsustainable and has irreversible adverse climatic impacts. An alternative is to use renewable biomass resources for fuels and chemicals. A biorefinery uses various types of renewable biological feedstocks to produce fuels and chemicals. Furthermore, sustainability of society is dependent on sustainable use of renewable resources including waste biomass. The book covers all the key topics relating to sustainable use of waste biomass resources along with biomass sources and characteristics, bioconversion technologies, processes for producing platform chemicals from biomass, and other biotransformations. Also, this book explains the process protocol for biochemical production. One chapter is dedicated to biomaterial production that can help in biorefinery processes. This book highlights new state-of-the-art aspects based on sustainable bioprocess technology involved in various sectors like industries, transportation, etc. This book will justify the needs of a majority of academicians and researchers around the globe in the field of biofuels and platform chemicals towards sustainability. I always accept suggestions and motivational inspirations from my parents for preparation of a manuscript. I express my sincere thanks to all my family members for providing me solid support and in time cooperation towards the completion of book manuscripts. I express our hearty thanks to Mr. Phil Carmical, Scrivener Publishing, LLC for his kind and timely help supplying full logistic support towards publication of this book.

Dr. Prakash Kumar Sarangi
Scientist, College of Agriculture Central Agricultural University, Imphal, Manipur, India

1

Biofuels: Classification, Conversion Technologies, Optimization Techniques and Applications

Sakthivel R[1]*, Abbhijith H[1], Harshini G V[1], Musunuri Shanmukha Vardhan[1] and Krushna Prasad Shadangi[2]

[1]Department of Mechanical Engineering,Amrita School of Engineering, Amrita Vishwa Vidyapeetham, Coimbatore, India
[2]Department of Chemical Engineering, Veer Surendra Sai University of Technology, Burla, Odisha, India

Abstract

To combat climate change, many researchers over the past few decades have focused their attention towards adopting and studying biofuels that could not just cause fewer toxic emissions compared to conventional petroleum-based fuel, but are also being generated from a myriad of plant waste, animal waste, etc. In this chapter, the primary study is focused on the classification of biofuels: 1st gen, 2nd gen, and 3rd gen biofuels. Then, different conversion technologies, such as pyrolysis, gasification, hydrothermal processes, and transesterification, which are commonly used in industries and labs to obtain biofuels from a plethora of biomass are studied in detail. After obtaining the biofuel from any one of the aforementioned conversion technologies, they could be used as an alternative power source and studied for emission and performance characteristics. However, to determine the optimal results, a suitable optimization technique would have to be employed. In this chapter, two such optimization techniques, namely Response Surface Methodology and Genetic Algorithm, are described in the view of optimizing engine parameters. Finally, a clear view is given into the application of biofuels in the transportation sector, particularly in the automotive and aviation sectors.

**Corresponding author*: sakthivelmts@gmail.com

Prakash Kumar Sarangi (ed.) Biorefinery Production of Fuels and Platform Chemicals, (1–30) © 2023 Scrivener Publishing LLC

Keywords: Biofuels, pyrolysis, gasification, hydrothermal processes, transesterification, response surface methodology, genetic algorithm

1.1 Introduction

Back in 2005, the petrol price in India was ₹ 59.29 per liter and in 2020 petrol prices are at an all-time high of ₹ 95.53 per liter. The hike in the gap of 15 years is about 62% and it has been concluded that there will be a hike in all fossil fuel-based products in forthcoming years. This trend can be seen all around the world since fossil fuels are being depleting at an alarming rate. Externality means a consequence of an industrial or commercial activity that affects other parties without being reflected in market prices. It is a well-known fact that the usage of fossil fuels causes pollution, but this factor of energy consumption and production of fossil fuels is not considered, hence, making this factor an externality. Under research done by the International Monetary Fund, there is an indication that economic and environmental costs due to fossil fuels add up to $5 trillion [1]. As of now, there is an unprecedented worldwide interest to reduce carbon dioxide emissions which is central to reducing greenhouse gas effects and improving the air quality of the metropolitan cities. In many countries around the globe, the quality of air has been extremely bad in spite of the preventive measures taken by the government. This can be accounted to the over usage of fossil fuels and emissions due to automobiles.

The report submitted by INERA (International Renewable Energy Agency) at the G20 summit to meet the central goals of the Paris climate change agreement showcased that clean energy can achieve up to 90% of energy-related carbon emissions. As of now, 24% of world power generation comes from renewable energy. The primary energy supply stands for energy production plus energy imports, minus energy exports, minus international bunkers, then plus or minus stock exchanges. Right now, in the 24% of available clean energy, 16% accounts for primary energy supply. To achieve desired decarbonization by the year 2050, renewable energy should account for up to 80% and 65% of the primary energy supply. Bringing down the carbon levels in the air can be achieved by only shifting to a clean energy source, or, in other terms, renewable energy. It is important and high time that we move on to adopt and advance renewable energy. Renewable energy is derived from a replenishable source such as the Sun (Solar energy), wind (Wind power), rivers (Hydroelectric power), hot springs (Geothermal power), tides (Tidal power), and biomass (Biofuels). Sunlight is the most abundant form of energy available to us and solar energy is currently constantly seeing a surge in its usage to generate electricity around the world. Solar Photovoltaic (PV) technology can proselytize sunlight to electricity with the use of PV materials. The prices of renewable energy technologies for electricity production in the solar and wind sectors have come down by 70% and 30%, respectively. Denmark has taken up the challenge of transition to 100% clean energy in which wind power is going to play a major role. INERA suggests that biofuel production should increase 10 times by 2050.

It would not be an understatement to say the way forward is renewable energy. Biomass is going to play a major role in the coming future. Biomass is plant or animal materials that can be processed to generate biofuel and such processes could also be used to generate heat and electricity. Biomass was the first method known to man to generate heat as early humans used wood logs to generate heat. There are many types of biomass; if we consider the type of production, it can be chemical or biological, liquid or gas if we consider the type, and heat, current, or transport if we consider the purpose [2]. Biomass is obtained from specific energy crops, agriculture crop residues, forest residue, processed wood residue, algae, municipal waste, etc. Figure 1.1 shows the various forms of biomass. Biofuels are derived from biomass which includes, but is not limited to, animal waste, plants, and algae. Three major benefits from biomass usage are greenhouse gas reduction, low dependence on foreign oil, and an increase of opportunities in forestry and agriculture fields. Ethanol and biodiesel are very well-known biofuels. Biofuels are distinguished into three categories, namely 1st gen. biofuel, 2nd gen. biofuel, and 3rd gen. biofuels. These biofuels can be obtained from different feedstock yields. The given table below lists various biofuel alternative feedstocks currently and long-term yields under the International Energy Agency (IEA).

1st generation biofuels are derived from biomass such as sugar, starch, and vegetable oils. To attain 1st generation biofuels, many well-known methods

Figure 1.1 Various forms of biomass.

such as fermentation, distillation, and transesterification are used. 2nd generation biofuels are processed through wood, organic, and food waste along with some specified biomass crops. 2nd generation biofuels biomass go through a pretreatment process in which lignin is broken down. This pretreatment consists of thermochemical or biochemical reactions. After the pretreatment, the process is parallel to the production of 1st generation biofuels. The 2nd generation biofuels generate higher energy yields compared to that of 1st generation biofuels. 3rd generation fuels specifically use algae as the feedstock to make biodiesel. The algae's oils are converted into biodiesel using a similar process as 1st generation biofuels. It is a well-known fact that 3rd generation fuels are highly energy-dense compared to 1st generation and 2nd generation biofuels. Unlike 1st and 2nd generation biofuels, 3rd generation biofuels do not depend on crops, which relieve the stress on water and land. Meanwhile, they can be termed as high-energy, renewable, and low-cost sources of energy. Seaweed is also being reviewed as a possible energy source for 3rd generation biofuels, with the end product being biomethane. The BMP (Biochemical Methane potential) values of seaweed stock vary from 101.7 ($L\ CH_4/kg\ VS^{-1}$) to 357.4 ($L\ CH_4/kg\ VS$) per the species selected [3]. In the last decade, there has been a lot of research done on fast pyrolysis that can be used to extract high bio-oil yields. In this method, raw biomass is rapidly subjected to heating under inert atmosphere and immediately condensed to obtained liquid product. Pyrolysis is accounted as thermal decomposition of the feedstock with a low level of oxygen [4]. Gasification is also one of the methods to obtain biofuels based on fermentation of the biomass to obtain products like ethanol, butanol, hydrogen, methane, and acetate. After gasification is completed, the obtained syngas is processed into acids and alcohols with the help of specific microorganisms by the fermentation process. Lignocellulosic biomass, after fermentation, goes through size reduction that can be achieved in two methods to obtain biofuels either by pretreatment, hydrolysis, fermentation, and purification or gasification and fermentation. Gasification takes place in a low oxygen environment and fuel abundant conditions with an equivalence ratio of 0.25 (mass of O_2/stochiometric mass of O_2) [5].

Hydrothermal processes are used for the extraction of third-generation biofuels which must be abstracted from microalgae and macroalgae. One of the main reasons for using hydrothermal processes for 3rd generation biofuels is due to high moisture content in the aquatic biomass. In this method, the biomass is processed wet in hot compressed water. This process is temperature dependent and we get different end products per the operating temperature of the operation. At lower temperatures (less than 200 °C) hydrothermal carbonization (HTC) occurs in which the end product is biochar. At intermediate temperatures (200 – 375 °C) and below the critical point, hydrothermal liquefaction (HTL) occurs giving biocrude as its end product. Finally, above the critical point (above 375 °C) a gasification reaction occurs with syngas as its primary product. The char produced from HTC can be used as manure and it helps to retain nitrogen and sulfur along with reducing the emissions. As known biocrude from HTL helps in upgrading the fuels and chemicals and coming to syngas from the hydrothermal gasification process, it can be either used for combustion or converted to hydrocarbons. In other words, hydrothermal processes stimulate the natural process which occurs in the fossil fuel reserves [6]. The oils obtained from biomass, which was used to run diesel engine cars at the Paris exposition in 1900, had a major flaw in that they were highly viscous. The oils were almost 10 to 20 times more viscous compared to the diesel fuel used. To encounter this problem transesterification was used. This is a chemical conversion process in which the oils are reduced to their corresponding fatty ester form, called biodiesel. This can be a great alternative to conventional fuels used for a compression ignition (CI) engine. A few advantages of biodiesel are they are non-toxic, biodegradable, and free from sulfur and carcinogenic compounds [7]. In the following sections, a detailed study on the above-mentioned processes is given.

1.2 Classification of Biofuels

1.2.1 First-Generation Biofuels

First-generation biofuels are known as conventional biofuels because they are made out of sugar, starch, corn, animal fats, or edible oil. Figure 1.2 shows the production process for 1st generation biofuels. They have a small negative impact on food society as the yield of biofuel is limited [8]. Production of these biofuels includes the processes of transesterification, distillation, and fermentation. The fermentation process includes starches and sugars. Fermentation of these produces primary ethanol, propanol, and butanol. Ethanol has 1/3 the density of energy of gasoline. Transesterification (see Figure 1.3) is the process of producing biodiesel by using plant oil or animal fats. This process occurs in the presence of the catalyst by mixing the plant oil or animal fats with alcohols (methanol). Distillation is the process of separating the main product from its by-products.

The main sources are corn, potatoes, sugarcane, vegetable oil, soybeans, and animal fat. Biodiesel, corn ethanol, and sugar alcohol are the major

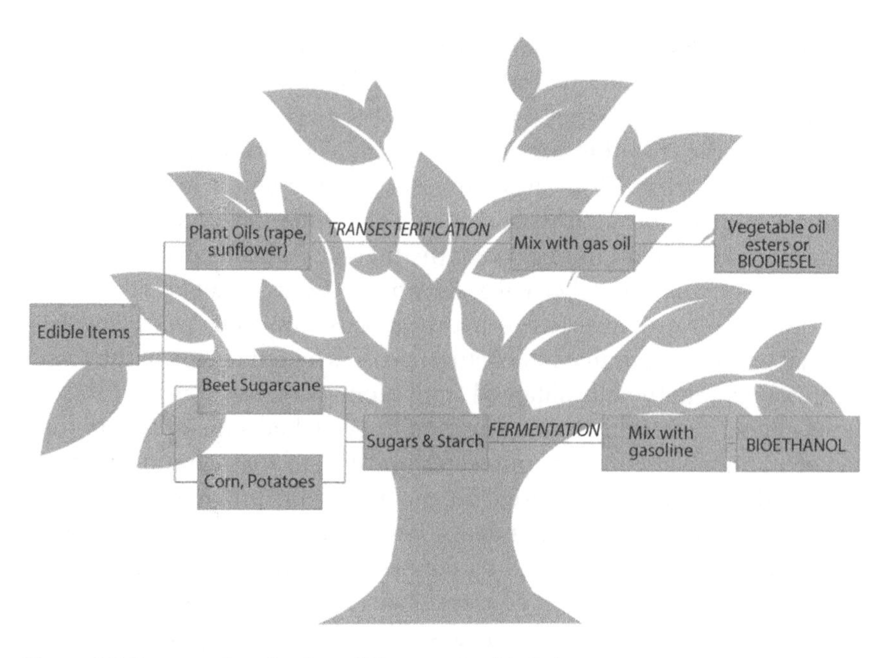

Figure 1.2 Process of production of 1st generation biofuels.

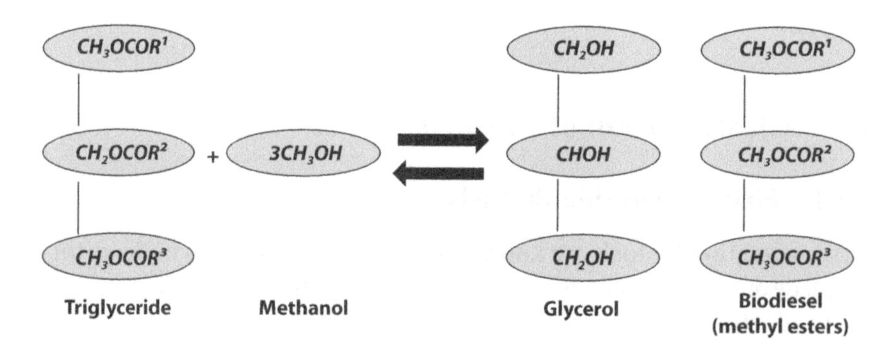

Figure 1.3 Transesterification process.

biofuels produced. A few problems which are faced are limited feedstock (food vs. fuel) and blending compatibility with conventional fuel, but there are a few benefits like environmentally friendliness, economics, and social security [9]. Table 1.1 shows the advantages and disadvantages of bioethanol. Table 1.2 shows the advantages and disadvantages of biodiesel.

The term sustainable development refers to the methodology that meets the demands of people without compromising the needs of the future. The three major pillars of sustainable development are society,

Table 1.1 Advantages and disadvantages of bioethanol [10].

Advantages	Disadvantages
Ethanol burns cleaner than gasoline	About 1.5 times of ethanol is required to produce the same amount of energy as gasoline
Produces fewer greenhouse gases	Corrosive to gasket rubbers
When blended, it increases the octane number of the fuel	Distribution through pipelines is difficult as ethanol absorbs water

Table 1.2 Advantages and disadvantages of biodiesel [10].

Advantages	Disadvantages
Biodiesel is less polluting than fossil diesel	Biodiesel is expensive, 1.5 times more than fossil diesel
Due to the lubricating properties, it increases the life of diesel engines and blending is easy with oils and other energy sources	Harms rubber hoses in diesel engines

economics, and the environment. Social acceptance, economic feasibility, and environmental impacts are the criteria on which the sustainability of a system is evaluated. To highlight the sustainability challenges, the life cycle of first-generation biofuel production systems can be analyzed. The production chain has several stages like site preparation, feedstock production, transportation, processing facilities, biofuel production, storage and dispensing, and combustion. The feedstock preparation is done with large-scale plantation for large-scale biofuel production of first-generation biofuels. Focussing on the environmental perspective, the benefits in terms of climate change mitigation have been under continuous sustainability discussion [10].

1.2.2 Second-Generation Biofuels

Second-generation biofuels are fuels derived from different feedstock, but especially non-edible lignocellulosic biomass. It is plant dry mass that is composed of cellulose and hemicellulose, which are carbohydrate polymers, and lignin, an aromatic polymer. Figure 1.4 shows the conversion

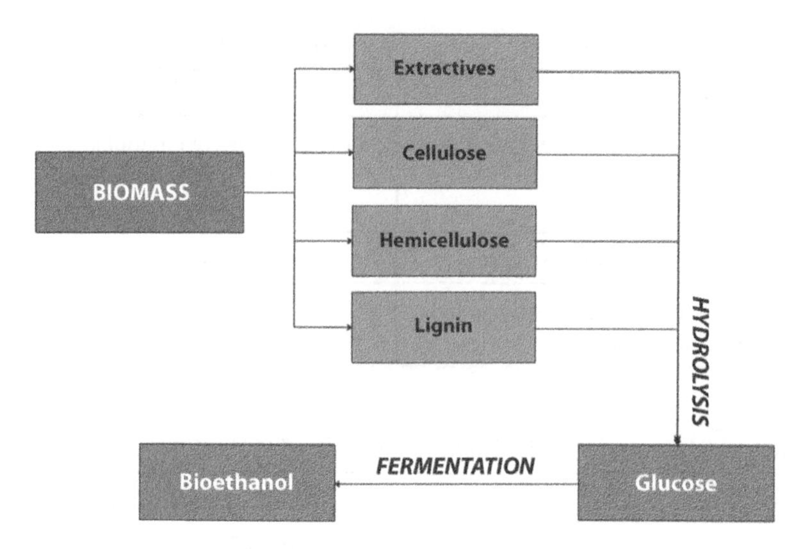

Figure 1.4 Conversion of lignocellulosic biomass to biofuel.

of lignocellulose biomass into biofuel. The production of these second-generation biofuels is classified into 3 main categories. They are homogeneous, quasi-homogeneous, and non-homogeneous. Homogeneous includes wood cuttings and white wood chips. Quasi-homogeneous includes agricultural and forest residues and non-homogeneous includes municipal solid wastes, which are low-value feedstock. The cost of this biomass is less than the cost of corn, sugarcane, vegetable oil, and other edible feedstocks [11]. The process of producing second-generation biofuels is more elaborate than first-generation biofuels as this requires pre-treating techniques to release the sugars which are trapped. This process requires more materials and energy. This biomass is more complex for conversion in general and the production is fully dependent on advanced technologies.

There are four steps in the lignocellulosic conversion process: (1) Pre-treatment, (2) Hydrolysis, (3) Fermentation, and (4) Distillation. The pre-treatment step is done to soften the biomass, thereby breaking down the cell structures. The hydrolysis process is done to increase the complexity of the sugar. The sugar is released from the cellulose part of the biomass. The fermentation process is done to convert these sugars to bioethanol. It is a metabolic process. Common crops used are salmon oil, jatropha, rubber tree, tobacco seed, jojoba oil, and sea mango. Biodiesel feedstocks include restaurant grease, non-edible oil crops, cooking oil waste, beef tallow, and pork lard. Figure 1.5 shows the feedstocks and products obtained in the hydrolysis process. Some of the advantages of second-generation

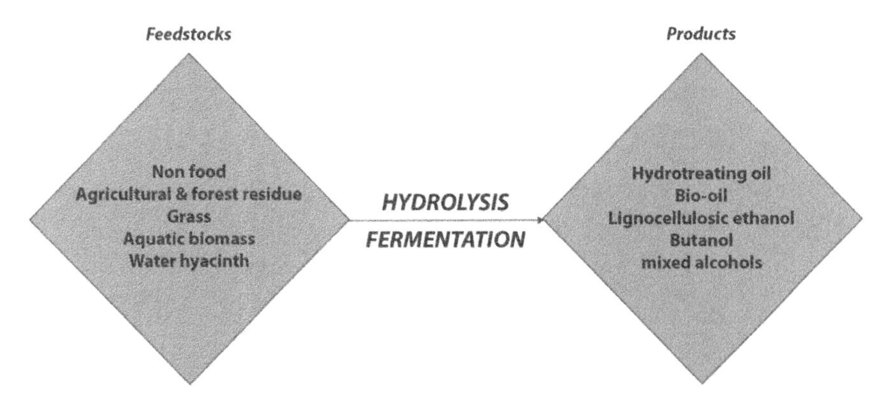

Figure 1.5 Feedstocks and products.

biofuels are it is environmentally friendly and does not compete with food items, so it does not affect people. One main disadvantage of these second-generation feedstocks is there are not enough active advanced technologies for the commercial usage of the waste generated by biodiesel production. Animal fats also contain a high concentration of saturated fatty acids. Those fatty acids increase the complexity of transesterification which is also a notable drawback [12].

In regards to constraints in the production of second-generation biofuels, water scarcity is one of the major problems. The biofuels produced from biomass have an increasing demand, which will intensify the pressure on clean water resources. The reasons for that are (1) certain feedstocks like energy crops are grown with a requirement of large quantities of water and (2) with crop production, agricultural drainage is likely to increase. When specific biomass crops are grown on large-scale land, it affects biodiversity, whereas production of biofuels from crops/forest residues should have less negative impact. Intensification of feedstock production has both positive and negative impacts on biodiversity [13].

1.2.3 Third-Generation Algal Biofuels

Third-generation fuels consist of microalgae, animal fat, fish oil, pyrolysis oil, etc. Third generation feedstocks have an upperhand compared to 1st and 2nd generation biofuels in terms of availability, economic feasibility, affecting of the food chain, and adaptability to climatic conditions. Figure 1.6 details the production of 3rd generation biofuels. Algae can be cultivated at lower costs due to their ability to grow in harsh conditions. Another advantage of the algae is the lipid content; on average the lipid content is about

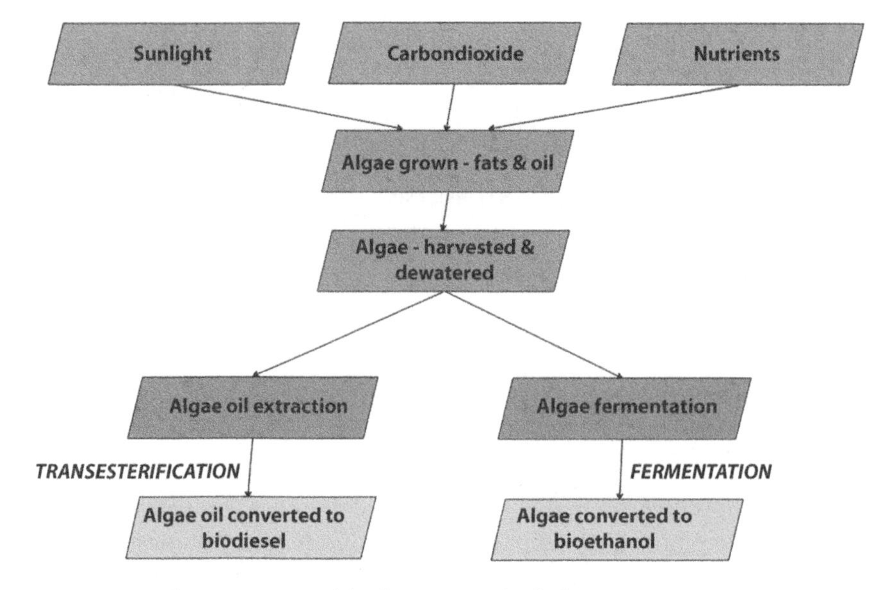

Figure 1.6 Production process of third-generation biofuels.

70% and by enhancing the conditions we can get up to 90% dry weight. Biodiesel can be extracted from waste cooking oil and is also known to be a cost-effective and heterogenous raw material. Increasing the usage of waste cooking oil may also reduce the burden on sewage treatment and water contamination [14].

Bioethanol from algal biomass is obtained by fermentation. In the process, strain selection and growth play a very important role in yield and property of the biofuel. Algal growth is very much dependent on environmental changes. Biodiesels are obtained by transesterification which are generally methyl or ethyl esters. The required fatty acid esters are obtained when triglycerides from oil feedstock and short-chain alcohol (methanol or ethanol) react along with a catalyst to introduce the alkyl group of the alcohol where glycerol is obtained as a side product [15].

1.3 Commonly Used Conversion Technologies

1.3.1 Gasification

Gasification has been in practice since the 1800s when, from coal, also known as town gas, gases were produced. In the 1900s, European nations

Table 1.3 Parameters and product description for different gasification technologies.

Technology → parameters & description↓	**Plasma gasification**	**Melting gasification**	**Fluidized bed gasification**	**Supercritical water gasification**	**Microwave gasification**	**Ref**
Temperature	>2500°C.	2200°C.	600°C.	375 – 500°C.	800°C.	[17]
Reactor type	Plasma gasifier	Fixed/Moving bed gasifier	Fluidized bed reactor.	Hydrothermal gasifier	Microwave-assisted	[17]
Product Description	There's a 100% carbon conversion. The syngas quality is very high.	The syngas obtained has to undergo a downstream treatment.	The syngas quality is high and it's relatively easy to control the operating parameters.	The gasification is carried out in the presence of a large volume of water. The methane in the syngas has to undergo further refinement.	The temperature profile of the syngas is uniform and better yield is observed when we use a catalyst such as Ni.	[17]

employed wood gasification to fuel their cars when there was a paucity of fuel [16]. Gasification is a thermochemical conversion where biomass, or any carbonaceous matter, solid or liquid for that matter, is converted to syngas. The major composition of syngas is CO and H_2. There is also substantial heat produced in the process with a temperature range of 600 - 1500°C [17]. An important point to note while doing experiments with gasification is that the oxygen feed must be less than the stoichiometric values since higher oxygen may lead to oxidation of products. For enhancing gasification, gasifying agents are often used such as steam, carbon dioxide, or oxygen. In addition to syngas, tar and char are also obtained. Tar is the liquid phase and char is the solid phase [18]. Before gasification, it is important that we pre-heat the feedstock biomass so that there is no moisture content in it.

Common gasification technologies used in the industry include plasma gasification, melting gasification, fluidized bed gasification, supercritical water gasification, and microwave gasification. Given below (Table 1.3) are the subsuming parameters and product description for the aforementioned technologies.

Given below (Figure 1.7) is the line diagram of the gasification process.

1.3.1.1 Factors Influencing Gasification

1. Particle Size

Smaller particle size biomass is preferred to larger sizes because, for the former, the time needed for the heat transfer to the center of the biomass particles from the walls of the reactor is less, therefore, enhancing the rate of chemical reactions. Also, when the particle size of the biomass is less, the H_2 concentration is high and the concentration of tar and char is less [18].

2. Moisture Content

The gasification efficiency significantly depends on the amount of moisture content in the biomass. The lower the moisture content, the higher

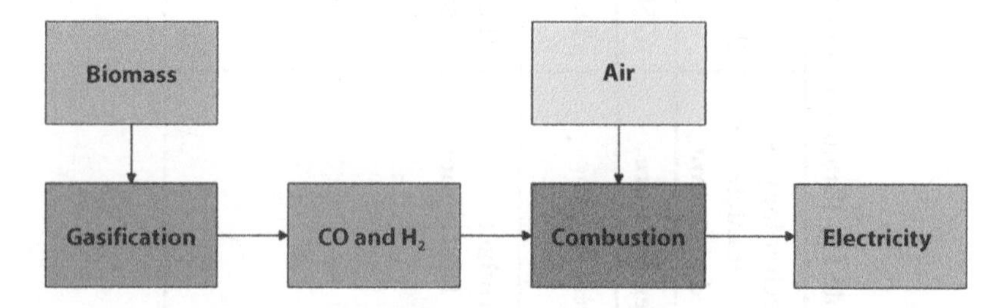

Figure 1.7 Line diagram of gasification.

the gasification efficiency. The favorable moisture content is 15% by weight moisture. Moisture content also affects the transportation and handling of biomass fuels. The problem with wet biomass is that some amount of energy is spent in vaporization and as a result of that, gasification temperature decreases [18].

3. Gasifying Agent
Gasifying agents are a vital medium for biomass gasification. The most commonly used gasifying agent is air. This is because it is cheap and is readily available. Despite these advantages, when air is used as a gasifying agent, in the synthesis gas predominantly N_2 is formed, which is not desirable. We could also use O_2 and steam as gasification agents but they are costly and have high operation costs [18].

4. Gasification Temperature
For reactions, high gasification temperatures are favored. This is because, at high temperatures, the decomposition of tar and char is high and the overall gas yield is also high. For exothermic reactions such as methanation and water gas shift reactions, lower temperatures are preferred [18].

5. Operating Pressure
The operating parameter is an important parameter because pressurized gasifiers generate gases of high pressure which are then fed into turbine coupled generators for the production of electricity. If the operating pressure is greater than the atmospheric pressure, we could store them in small volumes. The amount of heat transfer also increases when we used pressurized gasifiers. The disadvantages of pressurized gasifiers include complex construction, high operation costs, and difficulty in maintaining a constant flow rate of the biomass that is fed into the reactor [18].

1.3.2 Pyrolysis

Pyrolysis, also known as destructive distillation, is a thermochemical process where organic matter, in the absence of oxygen, undergoes decomposition. This process started in ancient Egypt for the production of tar. Later on, research of pyrolysis was expanded using a myriad of feedstock ranging from coal to wood to biomass. Biomass is the most common feedstock used these days in an industrial setting [19].

1.3.2.1 Production of Bio-Oil from Pyrolysis

First, the appropriate feedstock (rice husk, peanut shell, etc.) chosen for study must be ground into fine particles. Then, the particles must, using a screw feeder or a vibratory feeder, be fed into the reactor through the hopper. The mass flow rate of the biomass that is fed into the reactor is controlled using screw feed. It is also a good practice to pre-heat the biomass before it is fed into the reactor to remove the moisture content. To create an inert atmosphere inside the reactor, gases (usually N_2) are passed into the reactor. A heating coil is placed around the reactor core and it escalates the temperature of the reactor core. To improve the yield, in addition to the feedstock, catalysts such as Zirconium, Zinc, etc. can be used. Due to the high temperature in the reactor which is provided by the heating coil, the feedstock undergoes a phase change from the solid to the gaseous phase. The gases emanating are then passed through a freeboard where it expands and enters the cyclone separator. The separator takes advantage of gravity and allows the pyrolysis vapors to pass through and the solid char particles go down the spiral structure. The resultant pyrolysis vapors then enter a condenser, usually a shell and tube type, where it condenses to bio-oil. The following diagram (Figure 1.8) explains the pyrolysis process discussed.

Depending on the heating rate, the pyrolysis temperature, and the residence time of the biomass particles in the reactor, there are three types of

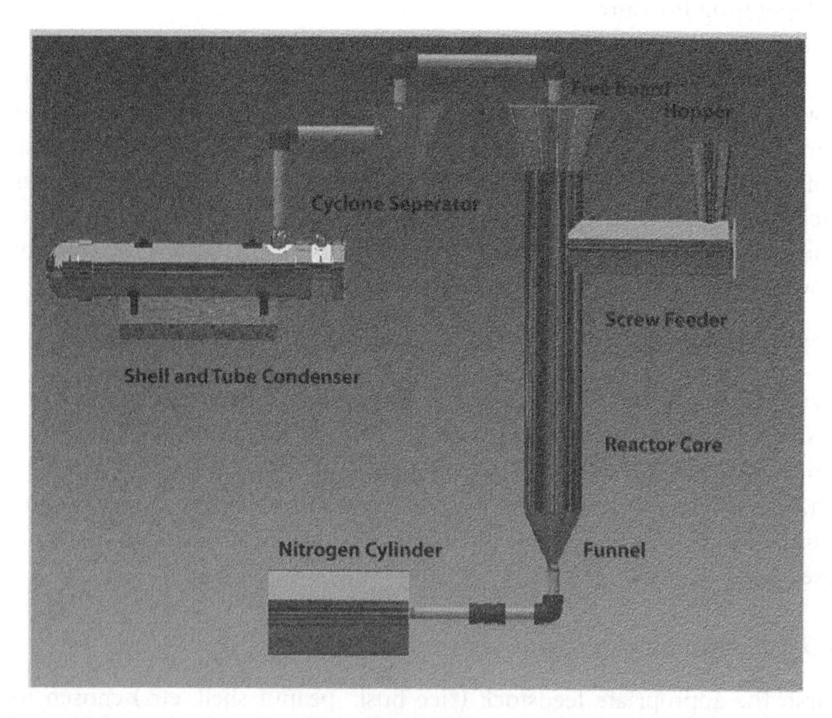

Figure 1.8 Setup diagram of pyrolysis reactor unit.

Table 1.4 Common parameters for different pyrolysis modes.

Parameters → Pyrolysis mode ↓	Temperature (°C)	Residence time	Heating value	Commonly used reactor
Slow Pyrolysis	300-550°C	5-30 minutes	10°C/min	Drum, Auger-type, and Rotatory kilns
Fast Pyrolysis	500-600°C	0.5-2 sec.	10-1000°C/s	Fluidized bed and microwave reactors
Flash Pyrolysis	900-1300°C	<0.5 sec.	>1000°C/s	Fluidized bed reactors

pyrolysis, namely: slow pyrolysis, fast pyrolysis, and flash pyrolysis. Table 1.4 shows the various parameters for the different pyrolysis modes mentioned above.

1.3.3 Hydrothermal Processes

These are the processes where an intricate organic waste, for instance lignocellulose biomass, is transmuted to simple organic compounds such as biochar, water, carbon dioxide, etc. The most commonly used type of wastes in hydrothermal processes are municipal solid wastes (MSW). This process occurs in an environment aided by high temperature and pressure where the organic wastes solubilize. To further aid the process, different oxidizing agents such as hydrogen peroxide (H_2O_2), are used. In the processes discussed previously, one of the key issues is that wet feedstock is not desirable and in case we are in need to use them, it is better to preheat the biomass before feeding it in the appropriate reactor. But, in hydrothermal processes, the processing of wet wastes can be achieved without the need for adopting energy-intensive dewatering steps. It is an advantage that hydrothermal processes have compared to thermochemical conversion techniques such as pyrolysis or gasification.

Based on the operating pressure and temperature, hydrothermal processes can be classified into three categories. They include a) Hydrothermal carbonization, b) Hydrothermal Liquefaction, and c) Hydrothermal gasification. Let's study each of the aforementioned processes in detail.

1.3.3.1 *Hydrothermal Carbonization*

Hydrothermal carbonization, also called wet torrefaction, is the process of converting organic solid wastes that are cellulose-rich, such as bagasse, straw, corn stover, plastics, etc., into useful products such as biochar. The energy density of the bio-char is very high. The operating conditions for hydrothermal carbonization are as follows: 180-250°C temperature and 2-10 MPa pressure. The reason for maintaining the low temperature is to prevent gasification or liquefaction so that only char (desired product) is produced. There are three steps in hydrothermal carbonization: decarboxylation, dehydration, and aromatization. In decarboxylation, carboxyl groups are removed and in dehydration the OH groups are removed. As a result of this, the oxygen/carbon ratio or O/C ratio, is reduced significantly and the final product has high energy density. No catalyst is needed for hydrothermal carbonization. The downside of hydrothermal carbonization includes heat loss, high residence time, unrestrained side reactions, etc. [20].

1.3.3.2 *Hydrothermal Liquefaction*

Hydrothermal Liquefaction (HTL) is a thermo-chemical process where wet biomass is converted into fuel. The feedstocks used in this process include, but are not limited to, algae and lignocellulose biomass. For woody biomass, it is advisable to reduce the particle size before the operation. The operating conditions for hydrothermal carbonization are as follows: 200-375°C temperature and 5-20 MPa pressure. There are three steps in hydrothermal liquefaction: depolymerization, decomposition, and recombination. During depolymerization, depending on the chemical and physical properties of the biomass feedstock, the macromolecules undergo a sequential dissolving process. In short, the long polymer chain gets broken down into a shorter hydrocarbon chain. In decomposition, oxygen is removed from the biomass and there is a formation of water and carbon dioxide. In recombination, the hydrogen fragments recombine and form a compound of high molecular weight, also known as coke [21]. Some advantages of HTL include a huge percentage of carbon recovery and high energy outputs [22].

1.3.3.3 *Hydrothermal Gasification*

Hydrothermal gasification is biomass gasification in hot compressed water. The water usage is two-pronged. First, the water is used as a solvent and second, it could be a reaction-aid. This is one of the most efficient technologies adopted for wet biomass gasification, as the reaction is short. The operating conditions for hydrothermal carbonization are as follows: 350-700°C temperature. The main products obtained include H_2, CO_2, and CH_4. Due to high operating temperatures, the decomposition is faster and complete. To take advantage of hydrothermal gasification, wet feedstocks are often used. Dry feedstocks could also be used, but for that we could adopt the regular gasification techniques discussed previously. Hydrothermal gasification usually involves the following steps: aqueous phase reforming, methane catalyzation, and supercritical water gasification. In aqueous phase reforming, biomass is mainly gasified to form hydrogen and carbon dioxide in the presence of a heterogeneous catalyst. In methane catalyzation, biomass is mainly gasified to form methane and carbon dioxide in the presence of a heterogeneous catalyst. The temperature range is between 350°C for methane production in the liquid phase and 400°C for production in the supercritical state. In supercritical water gasification, biomass is mainly gasified to form hydrogen and carbon dioxide without the presence of a solid catalyst [23]. The advantage of gasification includes not having tar in the product and a high heating value of the product.

1.3.4 Transesterification

Biodiesel is one of the most prominently used alternative fuels for diesel traction. Biodiesel is the fatty acid alkyl ester produced from vegetable or animal oils produced by transesterification reaction. Some of the common sources for biodiesel production explored by the current research community include Jatropha, canola, rapeseed, Calophyllum inophyllum, sunflower, castor, moringa, coconut, animal fat, pyrolysis oil, etc. [14]. The salient features of these feedstocks have been discussed in the previous topics. Transesterification is mainly carried out to convert raw fatty acid components of the feedstock into esters, using alcohol and base. This conversion ensured the reduced viscosity and enhanced combustion properties of the resulting biodiesel. During the reaction stage, the glycerides (mono, di, and tri) react with alcohol (preferably methanol) to produce esters and glycerol as main products. The alkyl group (R) alcohol is completely replaced by the alkyl ester (R') group. It should also be noted that oils with high free fatty acid (FFA>5) should be subjected to acid esterification and then followed by transesterification to avoid the formation of soap in the process. The overall transesterification reaction is consecutive stages, as shown in Scheme 1.1.

Scheme 1.1

$$TG + 3R\text{-}OH \longleftrightarrow GLY + 3\ FAME$$
$$TG + MeOH \longleftrightarrow DG + MET$$
$$DG + MeOH \longleftrightarrow MG + MET$$
$$MG + MeOH \longleftrightarrow GL + MET$$

where TG is Triglycerides, GLY is Glycerol, DG is Diglycerides, MG is Monoglycerides, MET is Methyl Esters, and FAME is Fatty Acid Methyl Esters.

In the course of the transesterification process, a base is commonly employed as a catalyst to fasten up the ester formation. The recent decade witnessed utilizing different types of catalyst to enhance the transesterification reaction for producing biodiesel. The catalyst can be classified as enzymatic, homogeneous, or heterogeneous based on their formulation and material. Bio-catalysis is yet another booming area in the field of biodiesel production, which used enzymes as a catalyst.

Moving on to the experimental conditions that affect the transesterification reaction, it was observed that the alcohol to methanol molar ratio, catalyst concentration, reaction temperature, and reaction time played a vital role in the yield of biodiesel. In general, varying the process conditions affects the yield and composition of biodiesel. The addition of catalyst to the reaction increases the conversion rate, to some extent, whereas excessive catalyst concentration has the potential to reverse the transesterification reaction. Meanwhile, high temperature and reaction time subdues the reaction rate and reduces the yield. The addition of methanol above the optimized point supports the formation of glycerol rather than biodiesel yield. Also, a higher methanol level degrades the rate of reaction owing to the physical dilution and flooding of active catalyst sites. So, it becomes inevitable to optimize the reaction conditions for the production of biodiesel to increase the yield which is given by Equation 1.1.

$$\text{Biodiesel yield (\%)} = \frac{\text{Mass of biodiesel generated (in g)}}{\text{Mass of raw oil utilized (in g)}} \times 100\% \tag{1.1}$$

Some of the experimental studies carried out to optimized the transesterification conditions to produce better biodiesel fuel are discussed. Raj *et al.* [24] modeled and optimized the reaction conditions for the production of biodiesel from Nannochloropsis salina with nanocatalysts derived from eggshell waste. The authors employed techniques like RSM and ANN for the optimization studies. The maximum FAME conversion was found to be 86% under the catalyst load of 3% (v/v), reaction temperature (60° C), time of 55 min, and oil to methanol ratio 1:6 (v/v). Rajendiran and Gurunathan [25] optimized the transesterification reaction parameters for biodiesel production from Calophyllum inophyllum oil. The zinc-doped CaO nanocatalyst was employed to enhance the yield and quality of biodiesel. A maximum conversion efficiency of 91.95% was recorded at 9.66:1 methanol to oil ratio, 5% catalyst load, 81.31 min reaction time, and 56.71°C reaction temperature. The overall green chemistry value has been evaluated as 0.873. Transesterification of soybean oil with ionic liquid catalyst is carried out by Panchal *et al.* [26]. The ANOVA results suggested 1:2 v/v of oil to methanol, 8% catalyst load, 4 h reaction time, and 300 rpm agitation as an optimized condition to obtain the maximum yield of biodiesel. The overall results of these optimization studies reveal that proper selection of reaction parameters not only affects the yield, but also the properties of biodiesel fuel produced.

1.4 Commonly Used Optimization Techniques

1.4.1 Response Surface Methodology

Response surface methodology (RSM) is one of the popularly known statistical tools which is gaining much importance in the renewable energy sector for optimizing energy production processes. It is a combination of mathematical techniques used in optimization and approximation of real-life stochastic models. RSM was presented by Box and Wilson in the early 1950s, commonly addressed as Box-Wilson methodology. RSM applies statistical techniques based on the factorial design aspects of central composite design (CCD) and Box-Behnken design (BBD). Apart from determining the optimized conditions from the least experimental trials, RSM also provides the information to evaluate the results of the experiments in the view of designing a process [27]. Due to this positive aspect, RSM is widely employed in the biofuel production process for optimizing particular responses associated with several variables. The response variable can be mathematically expressed by Equation 1.2 where the response 'y' depends on independent variables x_1 and x_2 with an error of 'e'.

$$Y = f(x_1, x_2) + e \tag{1.2}$$

In the recent optimization processes in the energy sector, both I and II order response surface models are utilized. The approximation of response of I order models are represented by a linear function having independent nature of variables. Equation 1.3 represents a simple I order model where β_0, β_1, and β_2 are regression coefficients.

$$Y = \beta_0 + \beta_1 x_1 + \beta_2 x_2 + e \tag{1.3}$$

On the other hand, the functions with 2 variables can be approximate3d by the II order model which also considers all cross-product and quadratic terms along with I order interaction models. A typical II order model can be mathematically expressed as Equation 1.4:

$$Y = \beta_0 + \beta_1 x_1 + \beta_1 x_2 + \beta_{11} x_1^2 + \beta_{22} x_2^2 + \beta_{12} x_1 x_2 + e \tag{1.4}$$

Meanwhile, the analysis of variance (ANOVA) is inevitable to evaluate the significance of models proposed by RSM. The P-value in ANOVA results plays a vital role to identify the model significance. In biofuel research, RSM plays a significant role in the optimization of conditions for fuel synthesis and engine parameters for fuel utilization. The optimization of the biofuel production process is already discussed in the previous section, due to which the engine parameter optimization is discussed here. Generally, engine operating parameters such as compression ratio, injection timing, load, and biofuel concentration are the most commonly optimized input parameters that respect output parameters such as thermal efficiency, fuel consumption, and emissions. The desirability function determines the optimal parameter in the multi-objective optimization problem. A typical representation of optimal parameter estimation from a multi-response problem is given in Figure 1.9.

Baranitharan *et al.* [28] optimized the engine attributed of DI-diesel engine fueled with Aegle marmelos pyrolysis oil blends using RSM and ANN. The authors observed that both the compression ratio (CR) and engine load affected the performance and emission attributes of the engine to a greater extent. The value of average correlation coefficient (R=0.998) and coefficient of determination (R^2=0.99) showed that the CR=17.5 and load=100% gave optimum engine operation for the selected fuel blend. Meanwhile, a similar engine study using Calophyllum inophyllum seed cake pyrolysis oil has been carried out by Sakthivel *et al.* [29]. In the research work, the authors optimized three different parameters, namely, CR, load,

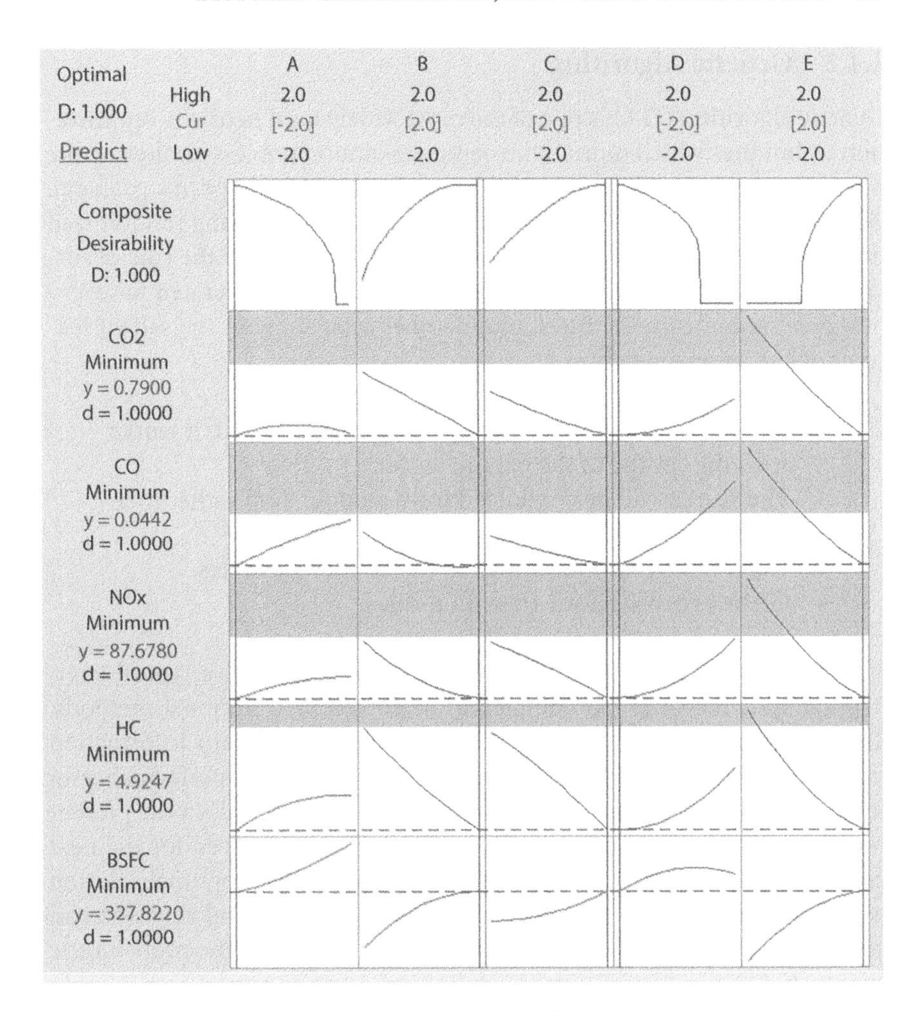

Figure 1.9 Multi-response optimization results using RSM.

and biofuel concentration. The CR and load directly affect the performance of the engine, whereas biofuel concentration affects the emission to a greater extent. The optimized values are found to be CR=18, blend ratio 20%, and load =100%. The reduction in NOx at exhaust with an increase in blend ratio [30] is attributed to lower combustion temperature by the authors [31]. On the other hand, Narayanan *et al.* [32] optimized NOx emission by varying injection holes, biodiesel blends, and loads. The variation in holes may vary wall impingement effects thereby reducing NOx. The maximum composite desirability of 0.715 was observed for input conditions of 6 kg load, 20% blend and 6 nozzle holes for optimum engine operation.

1.4.2 Genetic Algorithm

Genetic algorithm (GA) is one of the most widely used heuristic optimization techniques which mimics the natural evolution process by altering the population of an individual set of solutions required for a specific problem. GA operates on string structures that evolve with time using randomized but structured data exchange. The evolution takes place by the rule of survival of the fittest. Due to this, a new set of strings are generated for every new generation from the fittest old member sets. Some of the salient features of GA are given below:

- GA does not work directly on parameters, instead, it works with the coding of the parameter set.
- The search will be conducted from a population rather than a single point.
- GA utilizes payoff information rather than derivates.
- GA uses probabilistic transition rules.

GA is generally used for optimizing the parameters of a typical engineering system that is too complex to be solved by traditional methods. The advantage of GA comes from its capability to obtain information from previous solution sets and focused to enhance the performance of future solutions. GA works on selecting random individuals in the present population as parents and utilizes them to generate children for the next generation of solutions. By continuing the procedure, an optimal solution is attained since "good" parents create "good" children and then the bad points are rejected from the entire generation. Owing to the probabilistic approach, GA produces assorted solutions in different runs which leads to the requirement of multiple runs for deducing optimal solutions. The new population is usually generated by GA based on three important genetic operators, namely reproduction, crossover, and mutation.

The reproduction operator eliminates "bad" solutions from the population by duplicating the "good" solutions. Tournament and ranking selection are the two vital methods in reproduction operators. The prior one compared the two solutions and makes the best one survive, whereas the latter lists the entire population based on fitness values. Unlike reproduction, the crossover operator generates new solution sets by mating two parents to form a new child in a random mating pool. The "good" parent will always generate "good" children. The mutation operator is very much similar to the crossover operator in terms of generating a new population. The difference is that the new solutions will jump out from the local optimum. The optimization flow using GA is given in Figure 1.10.

Shirneshan *et al.* [33] optimized effects of biodiesel-ethanol blends in the operation of a diesel engine. The optimization is carried out using RSM and GA. The authors observed an increase in brake power by 30% with increased ethanol concentration in the blend. In the emission aspects, ethanol reduced smoke and NOx by 38% and 17%, respectively. The GA optimization results showed fuel blend (94.65%), speed (2800 RPM), and load (65.75%) as optimal conditions. Singh *et al.* [34] employed adaptive neuro-fuzzy interference system (ANFIS) along with ANFIS – GA method to optimize the engine operation fueled with Kusum biodiesel. Input variables such as FIP, FIT, blend, and load are optimized based on the engine performance and emission response. The results obtained from both the techniques were on par with each other. The ANFIS-GA method showed a more precise prediction of engine parameters as compared to ANFIS alone. The dual fuel combustion of diesel/gasoline in a CI engine was optimized by Xu *et al.* using computational fluid dynamics (CFD) and GA. The authors took seven inevitable operating parameters and optimized them in terms of fuel efficiency, NOx, and soot emissions. GA showed that about 45% of indicated thermal efficiency can be achieved by maintaining the emissions under Euro 6 level during the prescribed operating conditions [35]. Liu *et al.* [36] optimized operating parameters of a diesel engine

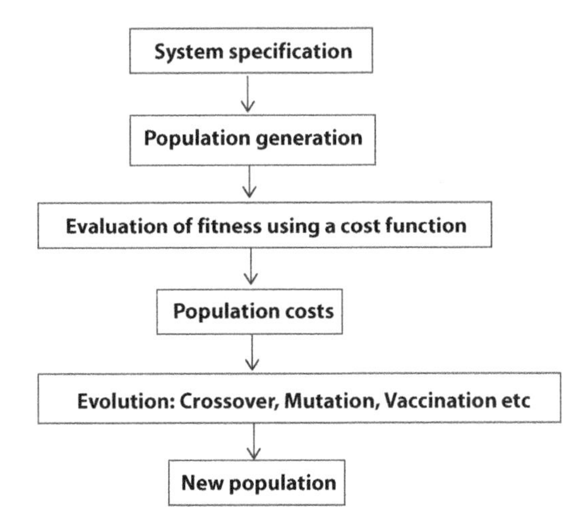

Figure 1.10 Optimization process using GA.

fueled with natural gas in dual fuel mode using GA. The optimization objectives selected in the study are ISFC, NOx, and methane emissions. From the results, it was observed that all the solutions meet soot limitations provided by Euro 6, whereas partial solution sets satisfied the NOx norms laid by Euro 5.

1.5 Application of Biofuels in Transportation Sector

1.5.1 Automobile Sector

As the demand for automobiles is constantly increasing, there is a huge threat of an increase of greenhouse gases accompanying it. Biofuels have a very promising application in the automobile sector. Using biofuels is not a new idea; in the 1990s ethanol was launched by blending it with conventional gasoline in two different concentrations, 10% and 85%, named E10 and E85 respectively. Interestingly, the Ford Model T (1908) was designed to run a mixture of gasoline and low level ethanol. Due to the increase of concern towards greenhouse gases, the prominence of the usage of biofuels has seen an exponential growth in the transportation sector. First-world countries already assessed their policies, costs, and outcomes for the usage of biofuels. Biofuel production has also been increasing constantly over the period. Biofuels can either be blended with fossil fuels for lower emissions or replace conventional fuels. Brazil, the European Union, and Nigeria are some examples of countries that made biofuels blending mandatory. Figure 1.11 shows the market value of biofuels projected from 2018 – 2026. Oxygenate is a fuel additive, generally alcohol or esters, which is rich in oxygen. Oxygenates reduce the carbon monoxide and soot produced during the combustion of the engine. The advantage of using E85, i.e., 85% ethanol and 15% gasoline, is that it can create more torque, can produce greater horsepower, and has a cooling effect on the engine. Biodiesel is also one of the biofuels used for blending. Biodiesel and conventional diesel have similar properties which make them suitable for blending. Furthermore, in the tests conducted using biodiesel using different engine test cycles, the average of these results pointed towards lower levels of carbon monoxide emissions, uncombusted hydrocarbons, and particulate matter released in comparison to conventional diesel [37]. In an experiment conducted using different percentage volume blends of pine oil, namely 20% (P20), 40% (P40) and 50% (P50), with diesel oil (P0) on a 4-cylinder CI engine built with a common rail injection system along with a turbocharged intake and

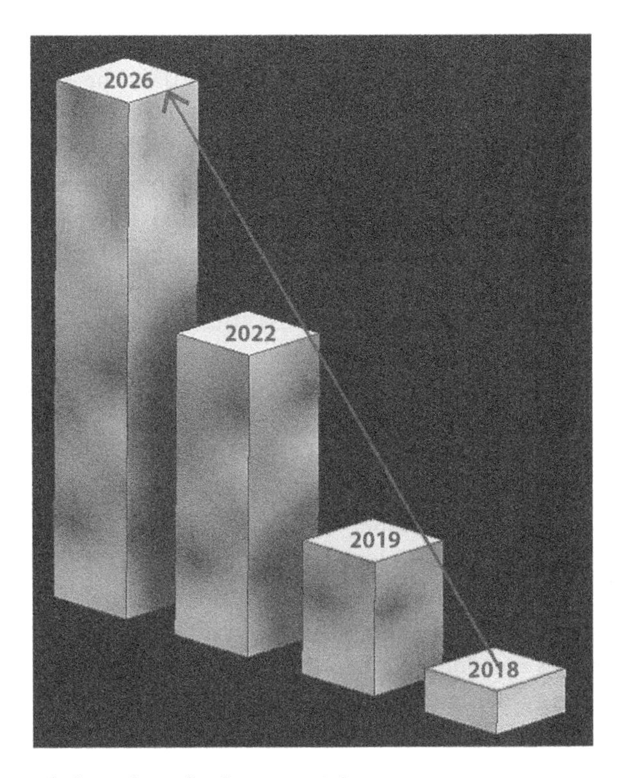

Figure 1.11 Biofuels market value (2018 – 2026).

mild exhaust gas unit. It was observed the energy release increased with better atomization provided due to better spray penetration of the pine oil blends which was achieved by an increase in the injection pressure. The P20 oil blend showcased a prominent decrease in emissions of soot and CO. The brake-specific fuel consumption of P20 and conventional diesel are quite close. As the blend level reaches 40%, there was a decrease in the favored characteristics. Brake horsepower (bhp) has been used as a definitive measurement of engine power and brake thermal efficiency is the ratio of the energy of the brake to the fuel energy [38].

1.5.2 Aviation Sector

To combat greenhouse gas emission in the aviation sector, instead of the conventional petroleum-based fuels that produce, upon burning, carbon dioxide, methane, and other greenhouse gases, biofuels could be used as a surrogate. Amongst all economic sectors, the aviation sector is growing expeditiously. This, in turn, leads to them releasing more toxic greenhouse gases into the atmosphere. Global aviation accounts for around 2% of global CO_2 emissions [39]. By 2050, the aviation industry has committed to reduce CO_2 emissions by 50% [40]. In 2008, for a flight from London to Amsterdam, Virgin Atlantic used biofuel instead of conventional aviation fuel in the ratio of 20:80 by volume as a test case scenario. Later, Air New Zealand, Continental Airlines, Japan Airlines, etc. started using biofuel blended with conventional fuels in different ratios [41]. Many countries such as Brazil, Germany, the USA, etc. have not only realized the urgency to address this issue, but have also come up with innovative solutions. The European Union (EU), since 2012, has executed the Emission Trading scheme for intra-European commercial flights to decrease carbon emissions [42]. Instead of using the conventional kerosene that is produced from the distillation of crude oil, eco-friendly kerosene that could fuel aircraft was produced by syngas obtained from biomass by Fisher-Tropsch synthesis. Different catalysts, such as K, Fe, Co, etc., could be used in the synthesis [43]. In Brazil, a novel approach was developed to enhance the sustainability of jet fuels that were produced from palm and soybean oil via life cycle assessment. First, the authors considered a baseline scenario where they studied via cumulative energy demand methods and conventional jet biofuel attributes, concerning the use of fossil hydrogen. In the second scenario, they investigated renewable surrogates for the production of hydrogen which can be achieved by biomass gasification, etc. They concluded that there was less carbon footprint for palm and soybean oil, an increase in life cycle energy efficiency, and drastic decreases in the range of 57 – 94% reduction in global warming impact [44]. One of the promising feedstocks for the production of jet-biofuel in Sao Paulo, Brazil is sugarcane. The authors studied the land availability for growing sugarcane and concluded that it is possible to increase the production of sugarcane by 73% within the mapped lands that are located within 25 km from processing units [40]. Biokerosene is an excellent sustainable fuel that could be used as aviation fuel. The different processes via which we can produce biokerosene include Alcohol-to-jet, Biogas-to-liquids, Biomass-to-liquids, and Hydro-processed Esters and fatty acids methods [45]. Although there is a myriad of pros to using aviation fuel obtained from biomass, we still have many challenges to overcome. For instance, the introduction of novel fuels would have to through go through the ASTM certification process, which is very time-consuming [46]. For a successful market launch of these fuels, it is not just about ensuring technological feasibility but also making sure that the cost is low, material availability is abundant, uncertainty in policies is being settled, etc. Thus, large-scale implementation of the aviation biofuel market is not completely achieved [39]. When studying the public attitudes towards the adoption of biofuels for aviation, the authors concluded that the public was not aware of the benefits of biofuels when they are used in aircraft and proposed that education and awareness campaigns must be conducted to make them cognizant of this prospect [41].

Conclusion

Fossil fuels are soon going to become "a thing of the past" because of their limited availability and high consumption rate, i.e., burning them daily for our needs for transportation, electricity, etc. They also release toxic greenhouse gases into the atmosphere, which is detrimental to not just human health, but to all forms of life. In this chapter, a broad study has been carried out on alternate renewable fuel, biofuel, one that is derived from biomass (plant or animal wastes). Compared to conventional petroleum-based fuels, biofuels are less flammable and have superior lubricating properties. The cost-benefit of adopting biofuels is also high and with a surge in demand, they could become an economical fuel in the future. Biofuels also ensure economic security as not all countries in the world have large reserves of crude oil and the economy of a country hits a dip when they are importing it from oil-rich nations. So, biofuels, that are being produced from locally grown crops in the host country would facilitate economic security.

Although biofuels' potential to bring down the carbon footprint is immense, more research has to be done to commercialize it, utilization of different feedstock and reducing the cost for its production, creating awareness about the merits of using them, etc. Biofuels not only help to combat climate change but can also bring many jobs to boost a country's economy.

References

1. Coady, D., *et al.*, *How Large Are Global Fossil Fuel Subsidies?* World Development, 2017. **91**: p. 11-27.
2. Donohue, T., *Welcome to Biofuels*. Biofuels, 2010. **1**(1): p. 1-2.
3. Allen, E., *et al.*, *What is the gross energy yield of third generation gaseous biofuel sourced from seaweed?* Energy, 2015. **81**: p. 352-360.

4. Patel, A., B. Agrawal, and B.R. Rawal, *Pyrolysis of biomass for efficient extraction of biofuel.* Energy Sources, Part A: Recovery, Utilization, and Environmental Effects, 2020. **42**(13): p. 1649-1661.
5. Slivka, R.M., M.S. Chinn, and A.M. Grunden, *Gasification and synthesis gas fermentation: an alternative route to biofuel production.* Biofuels, 2011. **2**(4): p. 405-419.
6. Biller, P. and A.B. Ross, *Hydrothermal processing of algal biomass for the production of biofuels and chemicals.* Biofuels, 2012. **3**(5): p. 603-623.
7. Demirbas, A., *Biodiesel Production via Rapid Transesterification.* Energy Sources, Part A: Recovery, Utilization, and Environmental Effects, 2008. **30**(19): p. 1830-1834.
8. Aro, E.M., *From first generation biofuels to advanced solar biofuels.* Ambio, 2016. **45 Suppl 1**(Suppl 1): p. S24-31.
9. Naik, S.N., *et al.*, *Production of first and second generation biofuels: A comprehensive review.* Renewable and Sustainable Energy Reviews, 2010. **14**(2): p. 578-597.
10. Naqvi, M. and J. Yan, *First-Generation Biofuels.* 2015. p. 1-18.
11. Lee, R. and J.-M. Lavoie, *From First- to Third-Generation Biofuels: Challenges of Producing a Commodity from a Biomass of Increasing Complexity.* Animal Frontiers, 2013. **3**: p. 6-11.
12. Alalwan, H., A. Alminshid, and H. Aljaafari, *Promising evolution of biofuel generations. Subject review.* Renewable Energy Focus, 2019. **28**: p. 127-139.
13. Carriquiry, M.A., X. Du, and G.R. Timilsina, *Second generation biofuels: Economics and policies.* Energy Policy, 2011. **39**(7): p. 4222-4234.
14. Sakthivel, R., *et al.*, *A review on the properties, performance and emission aspects of the third generation biodiesels.* Renewable and Sustainable Energy Reviews, 2018. **82**: p. 2970-2992.
15. Daroch, M., S. Geng, and G. Wang, *Recent advances in liquid biofuel production from algal feedstocks.* Applied Energy, 2013. **102**: p. 1371-1381.
16. Widjaya, E.R., *et al.*, *Gasification of non-woody biomass: A literature review.* Renewable and Sustainable Energy Reviews, 2018. **89**: p. 184-193.
17. Shahabuddin, M., *et al.*, *A review on the production of renewable aviation fuels from the gasification of biomass and residual wastes.* Bioresource Technology, 2020. **312**: p. 123596.
18. Motta, I.L., *et al.*, *Biomass gasification in fluidized beds: A review of biomass moisture content and operating pressure effects.* Renewable and Sustainable Energy Reviews, 2018. **94**: p. 998-1023.
19. Fahmy, T.Y.A., *et al.*, *Biomass pyrolysis: past, present, and future.* Environment, Development and Sustainability, 2020. **22**(1): p. 17-32.
20. Shen, Y., *A review on hydrothermal carbonization of biomass and plastic wastes to energy products.* Biomass and Bioenergy, 2020. **134**: p. 105479.
21. Gollakota, A.R.K., N. Kishore, and S. Gu, *A review on hydrothermal liquefaction of biomass.* Renewable and Sustainable Energy Reviews, 2018. **81**: p. 1378-1392.

22. Ponnusamy, V.K., *et al.*, *Review on sustainable production of biochar through hydrothermal liquefaction: Physico-chemical properties and applications.* Bioresource Technology, 2020. **310**: p. 123414.
23. Kruse, A., *Hydrothermal biomass gasification.* The Journal of Supercritical Fluids, 2009. **47**(3): p. 391-399.
24. Vinoth Arul Raj, J., *et al.*, *Modelling and process optimization for biodiesel production from Nannochloropsis salina using artificial neural network.* Bioresource Technology, 2021. **329**: p. 124872.
25. Naveenkumar, R. and G. Baskar, *Optimization and techno-economic analysis of biodiesel production from Calophyllum inophyllum oil using heterogeneous nanocatalyst.* Bioresource Technology, 2020. **315**: p. 123852.
26. Panchal, B., *et al.*, *Optimization of soybean oil transesterification using an ionic liquid and methanol for biodiesel synthesis.* Energy Reports, 2020. **6**: p. 20-27.
27. Sakthivel, R., *et al.*, *Prediction of performance and emission characteristics of diesel engine fuelled with waste biomass pyrolysis oil using response surface methodology.* Renewable Energy, 2019. **136**: p. 91-103.
28. Baranitharan, P., K. Ramesh, and R. Sakthivel, *Multi-attribute decision-making approach for Aegle marmelos pyrolysis process using TOPSIS and Grey Relational Analysis: Assessment of engine emissions through novel Infrared thermography.* Journal of Cleaner Production, 2019. **234**: p. 315-328.
29. Sakthivel, G., C.M. Sivaraja, and B.W. Ikua, *Prediction OF CI engine performance, emission and combustion parameters using fish oil as a biodiesel by fuzzy-GA.* Energy, 2019. **166**: p. 287-306.
30. Srihari, S. and S. Thirumalini, *Investigation on reduction of emission in PCCI-DI engine with biofuel blends.* Renewable Energy, 2017. **114**: p. 1232-1237.
31. Srihari, S., S. Thirumalini, and K. Prashanth, *An experimental study on the performance and emission characteristics of PCCI-DI engine fuelled with diethyl ether-biodiesel-diesel blends.* Renewable Energy, 2017. **107**: p. 440-447.
32. S, N., R. K, and S. R, *Optimization of nozzle hole number for a diesel engine fueled with kapok methyl ester blend.* Energy Sources, Part A: Recovery, Utilization, and Environmental Effects, 2019: p. 1-13.
33. Shirneshan, A., *et al.*, *Optimization and investigation the effects of using biodiesel-ethanol blends on the performance and emission characteristics of a diesel engine by genetic algorithm.* Fuel, 2021. **289**: p. 119753.
34. Singh, N.K., *et al.*, *Prediction of performance and emission parameters of Kusum biodiesel based diesel engine using neuro-fuzzy techniques combined with genetic algorithm.* Fuel, 2020. **280**: p. 118629.
35. Xu, G., *et al.*, *Computational optimization of the dual-mode dual-fuel concept through genetic algorithm at different engine loads.* Energy Conversion and Management, 2020. **208**: p. 112577.

36. Liu, J., B. Ma, and H. Zhao, *Combustion parameters optimization of a diesel/natural gas dual fuel engine using genetic algorithm.* Fuel, 2020. **260**: p. 116365.
37. Barr, M.R., R. Volpe, and R. Kandiyoti, *Liquid biofuels from food crops in transportation - A balance sheet of outcomes.* Chemical Engineering Science: X, 2021. **10**: p. 100090.
38. EdwinGeo, V., *et al.*, *Experimental analysis to reduce CO2 and other emissions of CRDI CI engine using low viscous biofuels.* Fuel, 2021. **283**: p. 118829.
39. Deane, J.P. and S. Pye, *Europe's ambition for biofuels in aviation - A strategic review of challenges and opportunities.* Energy Strategy Reviews, 2018. **20**: p. 1-5.
40. Martini, D.Z., *et al.*, *Land availability for sugarcane derived jet-biofuels in São Paulo—Brazil.* Land Use Policy, 2018. **70**: p. 256-262.
41. Filimonau, V., M. Mika, and R. Pawlusiński, *Public attitudes to biofuel use in aviation: Evidence from an emerging tourist market.* Journal of Cleaner Production, 2018. **172**: p. 3102-3110.
42. Chao, H., D.B. Agusdinata, and D.A. DeLaurentis, *The potential impacts of Emissions Trading Scheme and biofuel options to carbon emissions of U.S. airlines.* Energy Policy, 2019. **134**: p. 110993.
43. Martínez del Monte, D., *et al.*, *Effect of K, Co and Mo addition in Fe-based catalysts for aviation biofuels production by Fischer-Tropsch synthesis.* Fuel Processing Technology, 2019. **194**: p. 106102.
44. Vásquez, M.C., *et al.*, *Holistic approach for sustainability enhancing of hydrotreated aviation biofuels, through life cycle assessment: A Brazilian case study.* Journal of Cleaner Production, 2019. **237**: p. 117796.
45. Neuling, U. and M. Kaltschmitt, *Techno-economic and environmental analysis of aviation biofuels.* Fuel Processing Technology, 2018. **171**: p. 54-69.
46. Chiaramonti, D., *Sustainable Aviation Fuels: the challenge of decarbonization.* Energy Procedia, 2019. **158**: p. 1202-1207.

2

Technical Challenges and Prospects of Renewable Fuel Generation and Utilization at a Global Scale

Rajesh K. Srivastava

Department of Biotechnology, GIT, GITAM. (Deemed to be University), Visakhapatnam, India

Abstract

In current periods, renewable fuel generation and its utilization have become a big challenge for the world. There are many renewable energy resources reported and there are technical challenges noted for their generation at high efficiency at a lower cost. These challenges in developing and underdeveloped countries are greater with limited resources and technical skills, devices, or procedures. The massive quantity of fuel utilization finds itself in developed countries. This is due to the number of developmental activities and industrial growth. Another factor for fuel utilization is the exponential rate of population that needs fuel utilization for their daily needs. We are rich in resources of energy that require the utilization of fuel energy production, but it needs the advanced levels of technical approaches or devices for energy harnessing at a lower cost with high efficiency. We are continuously utilizing non-renewable energy like coal, natural gases, and petroleum oils that created the environmental degradation trends and natural health issues for the world from ancient periods. The authors will discuss biomass, the solar system, and wind resources for renewable energy generation and make it our habit to utilize this renewable energy to maintain a green environment.

***Keywords*:** Fuel energy, renewable sources, biological processes, technical approach, utilization, fuel generation

Email: rajeshksrivastava73@yahoo.co.in

Prakash Kumar Sarangi (ed.) Biorefinery Production of Fuels and Platform Chemicals, (31–58) © 2023 Scrivener Publishing LLC

2.1 Introduction

Fuel energy is stored in mobile vehicles and its energy density is reported to be important for the internal engine. Fuel energy densities are reported to various energy forms. The energy density of gasoline or diesel is reported as higher than all other forms (i.e., ethanol, methanol, and hydrogen fuel in gaseous and liquid forms or propene liquid). Further, the mass and volume of gasoline or diesel is also reported [1]. All these energy forms are cheap and found as primary energy. The values of alcohol forms for fuel energy are good and acceptable in Brazil. These are cheaper than gasoline or diesel prices [2]. Hydrogen fuel showed high apparent energy density ($12 MJkg^{-1}$), but it was not exploited due to the need of equipment to store at high weight and volume. Further, hydrogen fuel can be stored in pressurized tanks with more difficulty. Next, energy density of liquid hydrogen can be kept in cryogenic tanks and it is more promising for tasks for hydrogen storage due to its total mass. It has found that 40% of the energy contained in gaseous hydrogen is needed to liquefy the hydrogen [3].

The total energy balance is poor and some gases have shown tendencies for evaporation at slow rates even in very well insulted tanks. Further, metal hydrides forms storage can be done for gaseous hydrogen in passenger cars. The energy value of low temperature hydride titanium or iron is reported with high temperature hydride magnesium [4]. So, in practical operation this can be done with the combination of low and high temperatures to fulfill the hydrogen fuel storage. There are reports on energy density in forms of mass (7%) and volume (17%) for gasoline or diesel fuel. Further storage modes for electrical energy are known to contain very low energy density [5]. Currently, the lead traction battery available is reported to 0.2% of mass and (0.7%) volume of gasoline energy density. Advanced research work is focusing on higher capacity batteries and it has seen efforts in the development of sodium and sulfur batteries. Beta-aluminum oxide is reported to act as an ion conductor. In this arrangement, electrolytes can develop the best conductivity at 600K and can be heated initially at higher temperatures [4, 6].

Now, lots of research work is being done for feasibility of biofuel synthesis from water hyacinth mixed with cassava starch sediment. It applied biological and physical conversion approaches and later, this biofuel was utilized for the gross production process [7]. Bioconversion processes, like anaerobic digestion, are applied for biomethane production and pH 8.5, water hyacinth and cassava starch sediment (25: 75), C: N ratio (30), and hemophilic temperature (55°C) were applied as optimum conditions for maximum quantity methane (as biogas) production (436.8 ml CH_4. g COD^{-1}) and maximum chemical oxygen demand (COD) removal (87.4 %) [7, 8]. Next, physical processes were utilized for biobriquette that contained various ratios of water hyacinth and cassava starch sediment (such as 10:90, 20:80, 30:70, 40:60, or 50:50) for the best ratio of fuel properties [9].

These were very close to Thai community product standard with various heating values, 15.7 MJ/kg, 15.4 MJ/kg, 14.9 MJ/kg, and 14.6 MJ/kg were reported for 10:90, 20:80, 30:70, 40:60, and 50:50 water hyacinth and cassava starch sediment, respectively. Comparing physical and biological conversion processes, biological conversion processes were found to be more efficient for high quantity of gross electricity generation (3.90 kWh) [7, 9].

Microbial consortium design is reported as a newly emerging field for researcher tasks and tried to extend the biotechnology frontier for pure and mixed cultures. In this regard, microbial consortium has shown the ability of microbe uses for a broad range of carbon and nitrogen substrates or sources. Further, this approach can provide the microbes robust capability in environment stress/factors situations [10].

Microbes in a consortium are shown to perform complex functions that are not possible in single species of microbes. Further, uses of advanced technology can make it possible to understand the mechanism of microbial interactions via constructing consortia. Next, microbial consortium is reported to group modes of interactions and functions in their construction. Different trends in the study of microbial interaction and functions are reported and these are single-cell genomics (SCG), micro fluids, fluorescent imaging, or membrane separation [11]. A community profile study is done using polymerase chain reaction (PCR) denaturing gel electrophoresis (PCR-DGGE). Later, amplified ribosomal DNA restriction analysis (ARDRA) and terminal restriction fragment-length polymorphism (RFLP) are applied for these studies. Some examples are given for their possible application in area of biopolymers, bioenergy, biochemical productions, and bioremediation tasks [10, 12]. In this chapter, the author discusses the variation in biofuels synthesis and challenges, approaches, microbial systems and their application reported in bioprocesses.

2.2 Biofuel Synthesis

Now, fuel energy availability also reports big challenges for energy utilization for transportation purposes and a number of vehicle utilizations in the world now face big challenges to fulfill their energy demands with the ever increasing population. Further, fossil fuel is utilized in greater quantity for their vehicles' energy needs. It further generates a number of toxic gases and biproducts in our healthy atmosphere, causing health issues to biological objects likes humans, animals, and plants [13]. Sometimes, burning these fossil fuels in internal engines also creates problems for their future availability and continuous price rising that becomes a big challenge for many low and middle classes people with limited income sources [14]. It has been seen that non-renewable energy utilization is reported to produce carbon dioxide, methane, and other greenhouses gases (GHGs) in the atmosphere and their burning is continuously heating the planet, contributing to changing weather patterns, effecting the food production, animal ecosystems, and essential biodiversity within the habitat [13–15].

Next, a huge quantity of electricity supply is discussed in a sustainable manner with addressing a narrow frame of reference for setting up incremental decision making. Comparatively, the economic, social, and environmental aspects of renewable energy sources outweigh fossil fuels. Also, take into account the impact of the fossil fuel supply chain. Electrical generation from renewable energy was shown to be more economical compared to fossil fuel sources [16]. There are increased calls to incorporate the external cost of electricity generation into the price of electricity. For global warming issues, 85% CO_2 emission is reported from fossil fuel burning or combustion and fuel supply for oil and gas, coal, or nuclear source utilization. Next, the economic, environmental, and social consequences are reflected with historical and present fossil fuel supply chains and these have proven useful in avoiding limited reference frames for addressing the consequences of business-as-usual operations of the fossil fuel supply chain [17].

In India, it is reported that electricity generation is from coal burning resources and nearly 56% of electricity is reported from coal sources. Further, natural gas contributes 10% electricity generation. A number of climatic and environment issues are reported, including global warming, acidification, eutrophication, and ecotoxicity, as main problems along with carcinogens and respiratory organics or inmorganics generation with climatic changes [17, 18].

2.2.1 Biomass Energy

Biomass energy utilization is found by using wood, wood pellet, and charcoal for its direct burning for heating or cooking tasks or pretreatment for hydrolysis into monomers units for bioenergy development. This biomass energy can help in minimization of CO_2 emission via replacing fossil fuel utilization [19]. We are aware that wood can be harvested from forests or woodlots that need to thin or from urban trees that fall down or have to be cut down. As it is reported that wood smoke also contained harmful pollutants, such as CO (carbon monoxide) and particulate matter, modern burning stoves, pellets stoves, and fireplace insert are able to reduce the amount of particulate matter from burning wood [20]. In poor countries, wood and charcoal are used for cooking and heating fuel energy in major tasks. Now, people have shown trends to harvest the wood faster than trees can grow, which can promote more deforestation activity. Planting fast-growing trees for fuels and using fuel-efficient cooking stoves can help in slower deforestation via improving the environmental components [19, 21].

The impact of bioenergy energy has been studied with non-renewable energy consumption in a sustainable manner in OECD (Organization for Economic Co-operation and Development) countries. Information or data from 36 OECD members for the period of 1990 to 2017 has come with the GMM (Generalized Method of Moments) method. Further estimation results have shown the effect of biomass energy consumption per capita in a sustainable development manner. It has reported positive and statistical significance, but the effect of non-renewable energy development is found in a negative manner [22]. Based on this study, the use of biomass energy sources in poor countries caused the level per capita of sustainable modes to rise. It has found that OECD countries preferred to use biomass energy instead of non-renewable energy sources by reaching year 2030 SDGs (Sustainable Development Goals). So, biomass energy for sustainable development manners can show a more significant study [23].

In recent years, the globalization of fuel consumption was raised as a rapidly developing and influential aspect has for economic growth in developing countries. It has seen that developing countries can accelerate the pace of globalization with promotion of biofuel development at a growing scale or with ongoing globalization [24]. Many research papers have discussed the impact of globalization on biofuel in a panel data of 50 developing countries from the period of 2012 to 2016. Their panel estimation results showed economic globalization with positive effects in biofuel production. There is evidence shown in robust manners via checking of several robustness tasks, encouraging the economic aspect of globalization that can increase biofuel consumption to reduce the harmful environmental impacts [24, 25]. There are two forms of bioenergy, biofuel and biogases, that can be produced from biomasses of different kinds, such as microbial and plants [25].

2.2.2 Biofuels

In recent periods, researchers, as wells industrial people, have contributed to developing effective approaches for synthesis of biofuels and this has drawn increased attention at the worldwide level for substitution of petroleum derived fossil oil energy sources. Further, it addresses the increasing cost of fossil fuels and also helps in creating more energy security and reducing global warming issues or problems found with liquid fossil fuels. Biofuel can be applied to any liquid fuel energy made or synthesized from sugarcane, food grain products, or non-food materials. It can be like diesel and synthesized from soybean oil or any other plant oils including algal lipids contents [26].

Earlier, this fuel was developed as dimethyl ether or Fisher-Tropsch liquids (FTL) for utilization of lignocellulosic biomasses. Recent studies show for sugarcane ethanol demand to be in the range of 17.5 to 34.4 million tonnes (MT) in 2030. This biofuel demand can be completed without any new deforestation due to intensifying ranching practices and conversion of existing Brazilian cattle pastures for sugarcane crop. In Brazil,

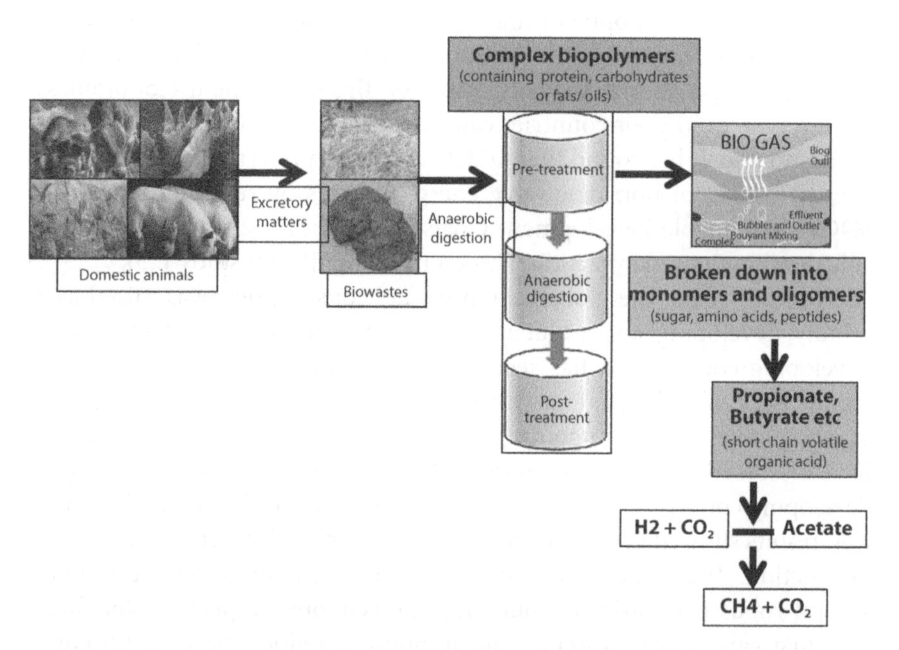

Figure 2.1 Biogases from anaerobic digestion of complex biopolymers from different wastes (domestic animals) used to enhance biofuel development.

cattle pastures for sugarcane crop production or cultivation is reported to expand up to five million hectares in 2030 for meeting the rising demands of ethanol biofuels from sugarcane crops. Figure 2.1 shows the biofuel development processes from biowastes [27].

In this regard, computer models are applied to compare the impact of economic, social, and policy scenarios on rising ethanol demand. Researchers have discussed the importance of broader strategic planning for sugarcane ethanol production in sustainable manner or modes. Watershed, municipal, and milled levels were taken into consideration in this waste matter utilization. There are five important multi-scale issues reported: the first point was the maintenance of long term water availability and quality, the second point was the enhancement of biodiversity and reversal of ecology based fragmentation [28], the third point was the planned elimination of sugarcane straw burning and subsequent enhancement in mechanical harvesting, the fourth point was a change in use of indirect and direct lands impacting capacity, and the last or fifth point was quality, availability, and durability of livelihood opportunities [27, 28].

Sugarcane crop is made from the *Saccharum* species and is found to belong to the Poaceae family. It is a perennial tropical grass plant used for sugar production in the world and the bagasse matter is also utilized for cellulosic ethanol production after efficient hydrolysis and pretreatment. Sugar juice was also used as resource for first generation ethanol or butanol production after use of effective yeast or bacterial species and can replace the conventional petroleum fuel energy, helping in the minimization of global warming. As we are aware that sugar is extracted from sugarcane stems, it was utilized for ethanol production. Further, its bagasse from sugar milling was pretreated and saccharified for cellulosic ethanol and butanol synthesis and used to generate steam and electricity [29].

The current total global level of renewable fuel is reported to be nearly 50 billion liters per year and sugarcane crop alone accounted for nearly 40%, showing that it is a major biofuel producing resource. In Brazil, conventional ethanol plants (CEP) using sugar was found to be effective, but the food debate for biofuel conversion was raised. Increased demand of cellulosic ethanol production using sugarcane bagasse was reported in biorefineries as a viable option for fossil fuels or first generation ethanol. Cellulosic ethanol is a more sustainable option than CEP [30]. In Brazil, energetic-environmental performance was reported in a biorefineries scenario and it compared the results of cellulosic production rates and costs against CEP and other alternative small scale systems. A multi-criteria approach, such as Embodied Energy Analysis, Ecological Rucksack, Energy Accounting and Gas Emission, was used for biofuel production at a global level [30, 31].

Published papers show tremendous success from sugarcane industries which are contributing to provide raw material for ethanol production. Brazil, especially, as well as other countries of the world have also shown interest for ethanol biofuel production utilizing cane juice or cane bagasse as a raw carbon or feedstock. A number of conventional technologies were reported as using sugarcane products or its waste matter containing fibers [32]. This matter was also used for various chemical synthesis. Genetic recombination was used for sugarcane novel biofuel synthesis in efficient ways. Many scientists have identified the key enzymes that actively contributed in the ethanol production processes can provide power in biorefineries and biofactories with tremendous potential [33].

These efforts have uplifted the socioeconomic status of a country in the world with a new route to sustainability of natural resources and it now provides more support for producing biofuels in developing countries such as India with the initiation of rural development efforts. Further, it can help create more job opportunities while saving and gaining more foreign currencies [32, 33].

Researchers have used first generation biofuel that was synthesized from utilization of food material or products including sugars, grains, or seeds with suitable microbial species, bioreactors, and process controls parameters [34]. In biofuel development, there are different feedstocks utilized as raw material or carbon substrates and it was found that sugars, grains, or seeds are the best feedstocks for first generation. This feedstocks utilization was dependent on hydrolysis steps such as pretreatment and schaachrification for developing the fermentating sugars utilized during the fermentation processes for effective and efficient microbial systems [34, 35].

As we are aware, a specific portion of above or ground level of raw materials are synthesized from plants such as potatoes or sugar beets (ground portion) and different types of fruits (as above portion of plant) can be used for development of biofuels or finish fuels utilizing simple processing [36]. In the last few decades, first generation biofuel development grew at a slower rate at commercial scale in different countries due to numbers of issues such as raising food price rates, insecure food product development, or poverty and hunger, so people are giving more emphasis to the development of second generation biofuel, normally made from non-edible lignocellulosic biomasses or non-edible food residues of crop production, including rice husks or corn stalks, or whole plant biomasses, such as grasses or trees, grown for energy development purposes [37]. Even today this type of biofuel development has not reached a commercial scale due

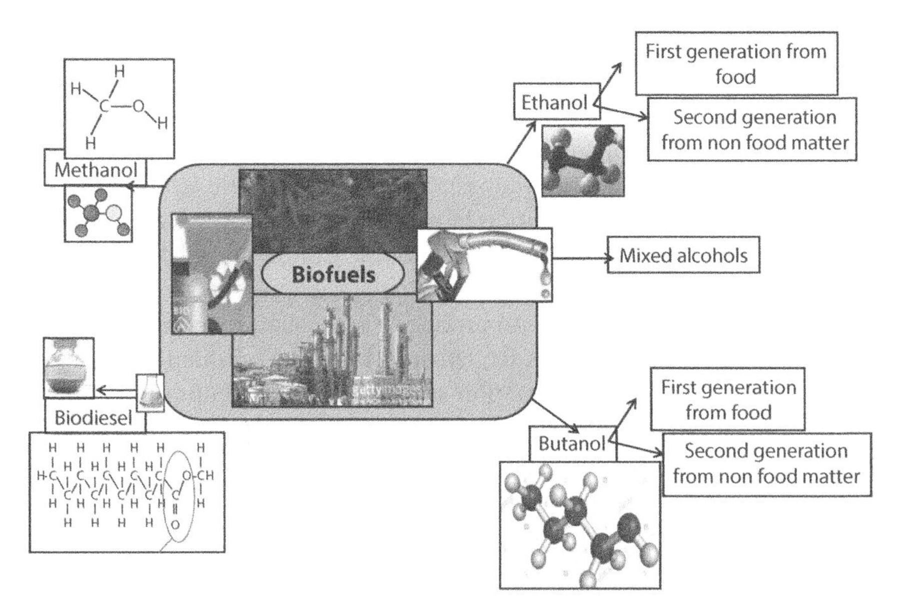

Figure 2.2 Different types of biofuels showing structures that are different generations. These have developed from different substrates via utilization of effective treatment and biofuel conversion processes with their own values and limitations.

to numbers of challenges such as effective microbial synthesis, pretreatment or saccharification processes, and optimal biological processes criteria in many countries in the world. Different types of biofuels are shown in Figure 2.2 [37, 39].

2.2.3 Biodiesel

There are different types of oil seed crops utilized for biodiesel synthesis or production and this biodiesel is kept under first generation biofuel. In this regard, Germany in 2005 showed its capability at a world level to use of rapseed and sunflower oil and nearly 2.3 billion liters were produced [39]. In Brazil, sugarcane crop cultivation rapidly increased nearly 7.6% per year from the last few decades. The state of Brazil (i.e., Sao Paulo in southern regions) showed its capability to support 55.3% of all Brazilian people cultivating sugarcane in 2010, utilizing their even more arable land values [38, 40].

Later, other parts of the world also started to produce at rapid rate including the United States (US). In the US, soybean was used to synthesize nearly 284 million liters of biodiesel in 2005 and 950 million in 2006, increasing 5% in 2013. In Brazil, the government has made the mandatory of blending 2% biodiesel into conventional diesel, stated from 2008, with a further increased to 5% in 2013 [41]. In 2008, 800 million liters of biodiesel was the goal from 590 million liters of biodiesel in 2005 in Brazil. This capacity (590 million liters of biodiesel in 2005) was derived from installed plants or capacity and was expected to double in 2008. In South East Asia, (Malaysia, Indonesia or Thailand), there is a palm biodiesel abundance due to utilization of palm oil for food uses [42]. Another plant, Jatropha (a non-edible oil tree), is now gaining more attention for production of biodiesel. This plant has shown capability to grow on varying qualities of land and is able to produce a huge quantity of its seeds that can form oils. Nowadays, in India, Jatropha biodiesel production is found as a part of a wasteland reclamation strategy and petroleum substitution or carbon emission reduction point of view shows more potential [43]. In India, Jatropha biodiesel is being pursued as part of a wasteland reclamation strategy. From the perspective of petroleum substitution or carbon emissions reductions this plant shows potential. Biodiesel derived from this plant's seeds is equivalent to starch based alcoholic fuels [43, 44].

The sustainability of diesel production from plants and their products raises questions regarding environmental issues in regards to more carbon emissions. Numbers of queries have been presented for the use of non-edible oils coming from *Jatropha curcas* L. and it is needed to promote palm oil utilization for a greater quantity of biodiesel synthesis. To complete the ever growing demand of energy fuel sources, it is needed to compare and validate the processes of biodiesel production using palm and jatropha oil products [45]. In this regard, the Life Cycle Assessment (LCA) approach was implemented to check the quality of biodiesel with the assessment of cultivation of the crops, the oil extraction stage, and, finally, the biodiesel synthesis stages. We found production of 1 tonne biodiesel from Jatropha plant products needs a land area nearly 118% higher than one tonne of palm biodiesel. So, the energy out to input ratio for palm oil needs is 2.27, which is slightly greater than Jatropha biodiesel (ratio~ 1.82) [46] and it was observed for carbon dioxide (CO_2) sequestration for a whole or complete life cycle chain of palm plant is twenty times higher than Jatropha originated biodiesel. All the results found superiority and sustainability in the nature of palm oils as the best feedstock in Malaysia [47]. It has been discussed in a number of literature that renewable forms of energy are the most promising energy when generated or synthesized from vegetable oils such as rapeseed, soybean, or palm oils, especially in Malaysia, as this country is known as world's second largest producer of palm oils [46, 47].

Malaysia has begun to align as the potential exporter country of palm oil to other parts of world as a promising feedstock for the food and oleochemical industries. Now, researchers are working on palm oil, Jatropha oil, and algal plant lipid contents for suitability as potential feedstocks for biodiesel yield and its quality for internal engines in Malaysia as an alternative and viable option for fuel energy. In this regard, many studies were done for utility of palm oil, Jatropha oil, and algal lipid content and found them suitable for biodiesel synthesis [48]. Now, there is a need to search for the best utility in plant products' lipids contents, but it is considered in its infancy stages. Malaysia is reported to implement B5 based palm oil as a suitable and the most abundant feedstock for biodiesel synthesis and it is proven as a safe fuel for engines without any engine modification. Table 2.1 shows the different types of biomass fuel energy [49].

Table 2.1 Bioenergy from different biomasses with different microbial synthesis.

Biofuel	Sources	Microbial systems	Reference
Second Generation Ethanol (4.7 to 8.7 g/L)	Whole stillage composed of undegraded bran and lignocellulose	Edible fungi *Neurospora intermedia* and *Aspergillus oryzae*	[67]
Ethanol Production Substrate (~52%, 55%, and 57%)	Green, brown, and red seaweed carbohydrate contents, respectively	Pichia. angophorae (CBS 5830)	[68]
Ethanol (~0.12 g/g and 0.09 g/g ethanol to dry seaweed, respectively)	Macroalgae *Ulva. rigida* with maximum total fermentable reducing sugar concentration of 34 mg/ml	Using adapted and non-adapted *Pachysolen tannophilus*	[69]
7% higher ethanol efficiency and improved ethanol yield	Hardwood spent sulfite liquor (HSSL) is rich in xylose, a monosaccharide that can be fermented into ethanol	*Scheffersomyces stipitis* C4	[70]

(*Continued*)

Table 2.1 Bioenergy from different biomasses with different microbial synthesis. (*Continued*)

Biofuel	Sources	Microbial systems	Reference
Optimum bioethanol yield and bioethanol titer (18.57 g/L) production, a green fuel	Fermenting sugar (98.7 g/L) from algae like Laminaria, Saccorhiza, Alaria and is capable of accumulating high starch/cellulose	*Saccharomyces cerevisiae* (ATCC 4126) with 15.09% of the volume immobilized yeast for 43.6 h with a 95% confidence interval	[71. 72]
Propanol (~4.5 g/L)	Utilized synergy of the native threonine pathway and the heterologous citramalate pathway for 2KB. 2KB is synthesized as intermediate in amino acid synthesis.	Pathway producing 2KB is constructed in engineered *E. coli*. This microbe converts biomass (i.e., glucose) to useful chemicals such as amino acids	[73]
Ethanol production (12.2% v/v) with dry matter production increased with fermentation time	From molasses (fermented) as well as by-products	Mutant *S. cerevisiae* with the viable cell count remained constant up to 50h of fermentation	[74]

(*Continued*)

Table 2.1 Bioenergy from different biomasses with different microbial synthesis. (*Continued*)

Biofuel	Sources	Microbial systems	Reference
High Butanol Titres (>2 g/L)	This strain utilized combined endogenous and exogenous pathways. This utilized α-ketovalerate as a supplement with yeast extract/peptone/dextrose (2% w/v).	Endogenous butanol synthesis in mutants *S. cerevisiae* ADH1	
Higher concentrations of solvent (12.42 g/L) and butanol (6.87 g/L), respectively at dilution rate of 0.02 h−1	Dilute Acid-Pretreated De-oiled rice bran production (42.12 g/L total sugars) that contained 25.57 g/L glucose, 15.1 g/L xylose, and 1.46 g/L cellobiose	*Clostridium acetobutylicum* YM1	[76]
Continuous packed-bed fermentation showed butanol productivity of 4.2 g/L h	Corn stover hydrolysate at a dilution rate of 0.44 h−1	Clostridium pasteurianum DSM 525 using suspended and surface immobilized cells on corn stover pieces	[77]

The above mentioned feedstock for biodiesel synthesis is reported as the best for domestic economic development by providing of job opportunities for many people for rural areas in the world including Malaysia. In addition, biodiesel added many advantages to the carbon emission rate in the world environment with more energy security compared to petroleum fuel energy [46]. Further, renewable fuel can reduce the toxic gases and particulate matter (PMs) in our healthy environment, creating the best conditions for biotic beings such as mankind. Many researchers have examined the emissions of most of internal engines that utilized the biofuel energy and found that palm oil and Jatropha feedstock, compared to algal plant products or oils, in Malaysia exhibited themselves as an alternative fuel [46, 48, 49].

2.3 Challenges for Bioenergy Generation

For the development of bioenergy, people (from industry to local level) are facing many challenges in regards to biomass supply from different sources, as well as operational, social, policy, and regulatory challenges at the local and commercial level. Further, biomass utility is reported for bioenergy synthesis as beneficial to meet the increased demand for fuel energy. Next, this biomass energy utilization is reported for reduction in carbon dioxide emission, global warming, and climatic change effects [50]. In the last few decades, people have developed an awareness for creating the biomass power industry at a rapid rate while accompanying some of these issue, so work needs to be done on the entire biomass supply chain so that it functions well with all tasks [51]. These are found to pose challenges for stable operation of enterprises with more impact on smooth development of the biomass power industry. Further, it showed a lack of optimization of biomass residue that can cause low utilization of biomass residues or feedstock. Most of industries are not aware this lacks benefit for that industry. This is why they are reluctant to take the risk for biomass utility for power generation [50, 51].

2.3.1 Operation Challenges in Biomass Energy Process

With operation challenges, in regards to bioenergy development, unavailability of sufficient feedstock is found as a critical challenge in biomass energy generation. This challenge is due to insufficient resource management. Government non-intervention approach was also found as a key factor to hinder the expansion or spreading of the biomass industry. Another big challenge is found in the regional and seasonal availability of biomass and proper storage issues or problems [52]. This is due to seasonal variations and can cause fuel price variation or fluctuation. Another issue is low energy density from biomass resources due to difficulty in acquisition of land for harvesting and storage. Ethyl alcohol or ethanol is found as the most promising liquid biofuel and in 2015, ethanol synthesis or biosynthesis reported nearly 25.6 billion gallons. The USA, Brazil, China, and European Union were involved with 28 other countries and set a target to blend ethanol with petroleum fuels [53]. In this regard, two major biosource matters, corn grains and sugarcane juice and its products, were utilized for ethanol biosynthesis that generated the first generation of ethanol. But, these two feedstocks were not sufficient to supply the current demand for alcohol synthesis. So, the next tasks were done with the utilization of non-edible lignocellulosic biomasses that were generated from multiple types of crops including corn and sugarcane [52, 53]. This helped in generation of second generation ethanol, butanol, propanol, and any other higher alcohol. Now, people have started to synthesize 3rd and 4th generation biofuels that can come from the utilization of algal biomasses or products like lipids. These types of biofuel could be beneficial for biorefinery products, utilizing biochemical and thermochemical approaches for biomasses utilization [54].

In regards to operational challenges, there is pressure on the transport sector due to moisture in biomass resources. Transportation of wet biomass was a big problem in production sites and caused energetic unfavorability with greater cost due to increased distance. Another challenge in regards to operational tasks is insufficiency of proper conversion facilities or plants with a shortage of equipment used in biofuel generation, pretreatment tasks, or saccharification [55]. Next, is the technical knowledge barrier that can result for non-standard qualities of bioenergy or equipment designs. As it is known that energy sources are found to be more diverse, they need appropriate pretreatment facilities, a requirement that can cause effective biodegradation with prevention of loss of heating value that can improve the production of biomass energy without increasing production costs and equipment investment. Table 2.2 shows the pretreatment processes for biofuel and reducing sugars [56].

In regards to the immature industry chain that is involved in biomass energy, it needs to minimize the virtually impossible tasks that occur in long term contracts for consistent feedstock supply at a reasonable price. Further, it is also needs to improve the process for high ability for gaining profits that come with many upstream firms and it needs to improve driving forces in the technology reform chain [55, 56].

Table 2.2 Biofuel synthesis with biomass hydrolysis challenges using pretreatment and microbial system.

Substrates	Pretreatment	Fermentation	Reference
Sugarcane bagasse (SCB) for renewable and abundant source for ethanol production	Chemical and physico-chemical pretreatment is applied to get best sugar yield, lignin removal, and cellulose content; Liquid hot water/ H2SO4 and ethanol/NaOH is common	Yeast based successful conversion of SCB to ethanol	[78]
Physical and chemical barriers due to lignin-carbohydrate composite in forest, agriculture, and agroindustrial residues	Successful pretreatment method increases the accessibility of holocellulose to enzymatic hydrolysis with the least inhibitory compounds being released for subsequent steps of enzymatic hydrolysis and fermentation	It helps in development of cellulosic ethanol and biorefineries via bacterial and yeast fermentation	[79]
Rice straw (RS) is an abundant, readily available agricultural waste	Pretreated with aqueous ammonia (27% w/w) at two pretreatment temperatures: room temperature and 60°C	The pretreatment release of glucose by enzymatic hydrolysis that is utilized for ethanol production	[80]

(*Continued*)

Table 2.2 Biofuel synthesis with biomass hydrolysis challenges using pretreatment and microbial system. (*Continued*)

Substrates	Pretreatment	Fermentation	Reference
Pretreatment of rye straw causes the disruption of lignin structures and breaks the linkage between lignin and the other carbohydrate fractions	Aqueous ammonia is applied for hydrolysis of this straw and the highest concentration of reducing sugars, chemical oxygen demand (COD), volatile fatty acid (VFA)	Conversion to fermentable sugars and is about 1.3 and 5.2 times higher at 60 °C and used in biofuel synthesis	[81]
Higher biogas yield from milled rice straw (1 mm)	1% NaOH pretreatment at ambient temperature shown more economical than the previous reports.	Produced 514 L biogas/kg VS (59% CH4) from rice straw	[82]
Maize straw is grown in corn crops worldwide including USA	Synergistic effect of alkaline pretreatment and Fe dosing (200–1000 mg/L) shown batch anaerobic digestion of maize straw and shortened TDT from 48 days to 13 days	Increased the methane yield 21.8% (368.8 mL CH4/gVS) and 56.2% (472.9 mL CH_4/gVS) at 4% NaOH and 6% NaOH, respectively	[83]

2.3.2 Economic Challenges in Biomass Energy Process

This type of challenge is due to feedstock acquisition cost for biomass resources. It has found that this is due to scattering of biomass resources, its cost of transportation (normally more expensive), and its need to reduce this transportation cost. Most of the biomass energy generating projects are eager to occupy the land close to biomass resources, leading to centralization of biomass energy generating projects that could help in energy security tasks [57].

Next, for contributing in biomass power generation plants, it is needed to evaluate the performance of plants in Portugal that can be dedicated for energy crops. This country, in regards to biomass power generation, has been analyzed for strategic, environmental, and economic interest in a project that is evaluated under the present renewable energy support (RES) support schemes [58]. The resulting values of the assumed Feed in Tariff (FIT) are not attractive points for private investors for biomass energy generation during these projects. We need to create a specific FIT scheme for this kind of biomass energy development that can do proper justification at perceived project risks by the expected strategic value of the environment via investment [57, 58].

Further challenges are reported for limited financing channels and high investment cost due to decentralization in capital, poor profitability, and frequent fluctuation of international crude oils prices with high market risks. Next, investors also seldom take initiative in the biomass power generation industry. Recently, biomass power generation is reported to have critical challenges due to excessive investment and high operating costs that are a big constraint in biomass energy. Improvement of the biomass pretreatment technologies that can minimize extra cost so that small scale fuel companies and scattered farmers can easily afford to supply their raw biomass is needed [59].

2.3.3 Social Challenges in Biomass Energy Processes

This area reports many challenges including conflicting decisions for land use issues and impact on the environment. These are detailed below.

2.3.3.1 Conflicting Decision on Utility of Biomass Resources

There is a problem in decision making on selecting of supplier, location, route, and technology development and these are in crucial need of proper communication. We need to work on strengthening leadership and implementing procedures of responsibility that are required for stakeholders. Further, they need to be fully aware of the economic, environmental, and social wealth or values in utilization of biomass resources [59, 60].

2.3.3.2 Land Use Issue or Problems on Biomass Cultivation or Utilization

The next tasks we need to work on are land use issues that are responsible for losses of ecosystem preservation and homes for indigenous people nearby [60].

2.3.3.3 Environmental Impact of Biomass Resources

It has been seen that biomass plantation can deplete the nutrient from soils and promote aesthetic degradation, as well as, to a great extent, losses of biodiversity on the earth. Other social issues or impacts can be seen from more installation of energy farms within rural areas or locations with enhancing of service opportunities and increased traffic issues. So, negative social impact has shown potential by ignoring benefits of new and permanent employment generation for people [61].

2.3.4 Policy and Regulatory Challenges for Biomass Energy Utility

Now, many countries have created policies for their nations and in the current period, the government is pushing the local people for biomass utilization by providing subsidizing to domestic fuel prices it has used for making the electricity for their location, generating the cost or benefit. It needs to lower of cost of conventional resources that can help in lowering the cost of power or fuel production from renewable resources [62, 63]. Further, this challenge can be minimized by using proper systems and can implement specific rules that regulate the work of utilization of biomass resources. It needs proper behavior with provisions minimizing the penalties when sometimes not used by some people. State, national, and international agencies are involved in regulation of special mechanisms to manage biomass resources development for biorefineries with implementation of relevant national and international standards and policies [64].

Further problems are found in supply of biomass at a large scale that can contain energy density and one of biggest challenges is found related to biomass moisture of conventional wood with 30% or higher percentage.

It enhances the transportation cost due to 300 kg water in one tone of wood. Further issues are biomass feedstock shape (chipped, pelletized, round, or baled form) that influenced the transportation economics [65]. To minimize this cost, it needs to supply compaction and densification of wood biomass while increasing efficient biomass supply [66].

2.4 Conclusions

This chapter discussed the fuel energy important for transportation tasks and other energy consuming processes worldwide. In this regard, lots of efforts has focused on generating fuel energy sources from natural resource utilization via utilization of suitable pretreatment or fermentative processes. We have faced problems with using non-renewable fuel energy like petroleum oils and natural gases that only started to degrade our environment due to toxic gases or particulate matter emission found to affect living beings including humans, animals, or plants. Other critical issues from non-renewable fuel energy consumption are its shortage of stocks in the future with higher prices in present or future periods. So, we need to start biomass energy research for more energy security for our nation and others. We face lots of challenges such as effective pretreatment, saccharification processes, and microbial processes for biofuel development. For second generation biofuel development, researchers are working to find out the effective pretreatment or saccharification processes with microbial fermentation that can produce cheap and huge quantities of biofuels. This chapter has focused on challenges such as operational, economic, social, and environmental issues that have stopped its commercial scale utilization. We need to promote utilization of local level resources with local people contributing for biofuel development. This chapter has highlighted it.

Abbreviations

ARDRA: Amplified Ribosomal DNA Restriction Analysis; **CEP:** Conventional Ethanol Plants; **CO**: Carbon Monoxide: **COD:** Chemical Oxygen Demand; **DGGE:** Denaturing Gel Electrophoresis; **FIT:** Feed in Tariff; **FTL:** Fisher-Tropsch Liquids; **GHGs:** Greenhouses Gases; **GMM**: Generalized Method of Moments; **kWh:** Kilowatt/hour; **MJkg^{-1}:** Mega Joule/ kilogram; **MT:** Million Tonnes; **LCA:** Life Cycle Assessment; **OECD**: Organization for Economic Co-operation and Development; **PCR:** Polymerase Chain Reaction; **PMs:** Particulate Matter; **RES:** Renewable

Energy Support; **RFLP:** Restriction Fragment-Length Polymorphism **SCG:** Single-Cell Genomics; **SDGs**: Sustainable Development Goals; **US:** United States

References

1. Afgan, N., Veziroglu, A., Sustainable resilience of hydrogen energy system. Int. J. Hydrogen Energ., 372013, 5461–5467, 2012.
2. Dincer, I.; Zamfirescu, C., Sustainable hydrogen production options and the role of IAHE. *Int. J.Hydrogen Energ.*, 37, 16266–16286, 2012.
3. Felseghi, R.-A., Carcadea, E., Raboaca, M. S., Trufin, C. N., Filote. C., Hydrogen Fuel Cell Technology for the Sustainable Future of Stationary Applications. *Energ.*, 12, 4593, 2019,
4. Azarikhah, P., Haghparast, S.J., Qasemian, A. Investigation on total and instantaneous energy balance of bio-alternative fuels on an SI internal combustion engine. *J Therm Anal Calorim.*, 137, 1681–1692, 2019.
5. Vinukumar, K., Azhagurajan, A., Vettivel, S.C., Vedaraman, N., Rice husk as nanoadditive in diesel–biodiesel fuel blends used in diesel engine. *J Therm Anal Calorim.*, 131(2):1333–43. 2018.
6. Wang, C., Yu, Y., Niu, J., Liu, Y., Bridges, D., Liu, X., Pooran, J., Zhang, Y., Hu, A., Recent Progress of Metal–Air Batteries—A Mini Review. *Appl. Sci.*, 9, 27872019.
7. Photong, N., Wongthanate, J., Biofuel production from bio-waste by biological and physical conversion processes. *Waste Manag Res.*, 38(1):69-77, 2020.
8. Zhang, Q, He, J, Tian, M., Mao, Z., Tang, L., Zhang, J., Zhang, H., Enhancement of methane production from cassava residues by biological pretreatment using a constructed microbial consortium. *Bioresour Technol.*, 102, 8899–8906, 2011.
9. Abbasi, T., Tauseef, S.M., Abbasi, S.A., Anaerobic digestion for global warming control and energy generation: An overview. *Renew. Sustain. Ener. Rev.*, 16, 3228–3242, 2012.
10. Bhatia, S.K., Bhatia, R.K., Choi, Y.-K.,Kan, E., Kim, Y.-G., Yang Y.-H., Biotechnological potential of microbial consortia and future perspectives. *Crit. Rev. Biotechnol.*, 38(8):1209-1229, 2018.
11. Johns, N.I., Blazejewski, T., Gomes, A.L., Wang, H.H., Principles for designing synthetic microbial communities. *Curr Opin Microbiol.*, 31,146-153, 2016.
12. Niu, B., Wang, W., Yuan, Z., Sederoff, R.R., Sederoff, H., Chiang, V.L., Borriss, R., Microbial Interactions Within Multiple-Strain Biological Control Agents Impact Soil-Borne Plant Disease. *Front Microbiol.*, 9, 11:585404. 2020. https://doi.org/10.3389/ fmicb.2020. 585404

13. Perrons, R.K., Richards M.G., Applying maintenance strategies from the space and satellite sector to the upstream oil and gas industry: A research agenda. *Energ. Polic.*, 61, 60–64, 2013. https://doi.org/10.1016/j.enpol.2013.05.081
14. Prosser, T., Financialization and the reform of European industrial relations systems. *Europ. J. Industr.Relat.*, 20(4), 351–365, 2014. https://doi.org/10.1177/0959680113505178
15. Olson, C., Lenzmann, F., The social and economic consequences of the fossil fuel supply chain. *MRS Energ. Sustainabil.*, 3, 6, 2016. DOI: https://doi.org/10.1557/mre.2016.7[Opens in a new window]
16. Hansen, J., Sato, M., Hearty, P., Ruedy, R., Kelley, M., Masson-Delmotte, V., Russell, G., Tselioudis, G., Cao, J., Rignot, E., Velicogna, I., Kandiano, E., von Schuckmann, K., Kharecha, P., Legrande, A.N., Bauer, M., Lo, K-W., Ice melt, sea level rise and superstorms: Evidence from paleoclimate data, climate modeling, and modern observations that 2 C global warming is highly dangerous. *Atmos. Chem. Phys. Discuss.*, 15, 20059–20179, 2016. doi:10.5194/acp-16-3761-2016
17. Friedlingstein, P., Andrew, P., Rogelj, R.M., Peters, J., Canadell, G.P., Persistent growth of CO2 emissions and implications for reaching climate targets. *Nat. Geosci.*, 7, 709–715, 2014. https://doi.org/10.1038/ngeo2248
18. Heede, R., Tracing anthropogenic carbon dioxide and methane emissions to fossil fuel and cement producers, 1854–2010. *Climatic Chang.*, 122, 229–241, 2014. https://doi.org/ 10. 1007/s10584-013-0986-y
19. Güney, T., Kantar, K., Biomass energy consumption and sustainable development. *Internat. J. Sustain. Developm. World Ecol.*, 27(8), 762-767, 2020. https://doi.org/10.1080/13504509. 2020.1753124
20. Alola, A.A., Yalçıner, K., Alola, U.V., Akadiri, S.S., The role of renewable energy, immigration and real income in environmental sustainability target. Evidence from Europe largest states. *Sci Total Environ.*, 674, 307–315, 2019. https://doi.org/10.1016 /j.scitotenv. 2019.04.163
21. Apergis, N., Danuletiu, D.C., Renewable energy and economic growth: evidence from the sign of panel long-run causality. *Int J Energy Econ Policy.*, 4(4):578–587, 2014. https://www.econjournals.com/index.php/ijeep/article/view/879/515
22. Aydın, M., The effect of biomass energy consumption on economic growth in BRICS countries: A country-specific panel data analysis. *Renew. Ener.*, 138:620–627, 2019. https://doi.org/10.1016/j.renene.2019.02.001
23. Bilgili, F., Koçak, E., Bulut, U., Kuşkaya, S., Can biomass energy be an efficient policy tool for sustainable development? *Renewable Sustainable Energy Rev.*, 71, 830–845, 2017. https://doi.org/10.1016/j.rser.2016.12.109
24. Subramaniam, Y., Masron, T.A., The impact of economic globalization on biofuel in developing countries. *Ener.Convers. Manag:*.X, 100064 (online 29 December 2020). https://doi.org/10.1016/j.ecmx.2020.100064

25. Bildiricil, M.E., Özaksoy, F., The relationship between economic growth and biomass energy consumption in some European countries. *J. Renew. Sustain Ener.*, 5, 023141 (2013); https://doi.org/10.1063/1.4802944
26. Talukdar D., Verma D.K., Malik K., Mohapatra B., Yulianto R., Sugarcane as a Potential Biofuel Crop. In: Mohan C. (eds) Sugarcane Biotechnology: Challenges and Prospects. 2017, pp. 123-137. Springer, Cham. https://doi.org/10.1007/978-3-319-58946-6_9
27. Taherzadeh, M.J., Lennartsson, P.R., Teichert, O., Nordholm, H., Bioethanol production processes. In: Babu V, Thapliyal A, Patel GK (eds) Biofuels production., Beverly, pp 211–253, Scrivener Publishing, 2013
28. Duarte, C.G., Gaudreau, K., Gibson, R.B., Malheiros. T.F., Sustainability assessment of sugarcane-ethanol production in Brazil: A case study of a sugarcane mill in São Paulo state. Ecological Indicators, 30, 119-129, 2013, Front. Energy Res. 5:7. https://doi.org/10.1016/j .ecolind. 2013.02.011
29. Agostinho, F., Ortega., E., Energetic-environmental assessment of a scenario for Brazilian cellulosic ethanol. *J.Cleaner Product.*, 47, 474-489, 2013, https://doi.org/10.1016 /j.j clepro. 2012.05.025
30. Martinelli, L.A., Filoso, S., Expansion of Sugarcane Ethanol Production in Brazil: Environmental and Social Challenges. *Ecologic. Applicat.*, 18(4) 885-898, 2008. https://www. jstor.org/stable/40062197.
31. Reis C.E.R., Hu B., Vinasse from Sugarcane Ethanol Production: Better Treatment or Better Utilization?. *Front. Energy Res.* 5, 7, 2017. | https://doi.org/10.3389/fenrg.2017.00007
32. Ojeda, K., Avila, O,, Suarez J., Kafarov, V., Evaluation of technological alternatives for process integration of sugarcane bagasse for sustainable biofuels production—part 1. *Chem Eng Res Des* 89(3):270–279, 2011. https://doi.org/10.1016/j.cherd.2010.07.007
33. Patrick, J.W., Botha, F.C., Birch, R.G. , Metabolic engineering of sugars and sugar derivatives in plants. *Plant Biotechnol* J., 11, 142–156, 2013.
34. Limayem, A., Ricke, S.C., Lignocellulosic biomass for bioethanol production: Current perspectives, potential issues and future prospects. *Progress in Ener. Combus.* Sci., 38(4), 449-467, 2012. DOI: 10.1016/j.pecs.2012.03.002
35. Cheng, K.-K., Cai, B.-Y., Zhang, J.-A., Ling, H.-Z., Zhou, Y.-J., Ge, J.-P., and Xu, J.-M. (). "Sugarcane bagasse hemicellulose hydrolysate for ethanol production by acid recovery process. *Biochem. Eng. J.* 38(1), 105-109, 2008. DOI: 10.1016/j.bej.2007.07.012
36. Wargacki, A. J., Leonard, E., Win, M. N., Regitsky, D. D., Santos, C. N. S., Kim, P. B., Cooper, S.R., Raisner, A., Herman, R.M., Sivitzl, A.B., Lakshmanaswamy, A., Kashiyama, Y., Baker, D., Yoshikun, Y., An engineered microbial platform for direct biofuel production from brown macroalgae. *Sci.*, 335(6066), 308-313, 2012. DOI: 10.1126/science.1214547
37. Xu, J., Li, M., and Ni, T., Feedstock for Bioethanol Production from a Technological Paradigm Perspective. *BioRes.* 10(3), 6285-6304, 2015. DOI: 10.15376/biores.10.3.Xu.

38. Rocha, G.J.M., Goncalves, A.R., Oliveira, B.R., Olivares, E.G., Rossell CEV (2012) Steam explosion pretreatment reproduction and alkaline delignification reactions performed on a pilot scale with sugarcane bagasse for bioethanol production. *Ind Crop Prod* 35(1), 274–279. https://doi.org/10.1016/j.indcrop.2011.07.010
39. Meyer, P.M., Rodrigues, P.H.M., Millen, D.D., Impact of biofuel production in Brazil on the economy, agriculture, and the environment. *Animal Front.*, 3(2), 28–37, 2013. https://doi.org/10.2527/af.2013-0012
40. Lee R., Lavoie J.M., From first to third generation biofuels: Challenges of producing a commodity from a biomass of increasing complexity. *Anim. Front.*, 3, 6–11, 2013. https://doi.org/10.2527/af.2013-0010
41. Liang, Y., Sarkany, N. Cui, Y. Biomass and lipid productivities of *Chlorella vulgaris* under autotrophic, heterotrophic and mixotrophic growth conditions. *Biotechnol Lett.*, 31(7):1043-9, 2009. doi: 10.1007/s10529-009-9975-7.
42. Mirzaie, M.M.A., Kalbasi, M., Mousavi, S.M., Ghobadian, B., Investigation of mixotrophic, heterotrophic, and autotrophic growth of *Chlorella vulgaris* under agricultural waste medium. *Prep Biochem Biotechnol.*, 46(2), 150-6, 2016. doi: 10.1080/10826068.2014.995812.
43. Gmünder, S., Singh, R., Pfister, S., Adheloya, A., Zah, R., Environmental Impacts of Jatropha curcas Biodiesel in India. *BioMed Resear. Internat.*, 2012, 623070, 10, 2012. https://doi.org/10.1155/2012/623070.
44. Lam, M.K., Lee, K.T. Rahmanmohamed, A., Life cycle assessment for the production of biodiesel: a case study in Malaysia for palm oil versus jatropha oil. *Biofuels, Bioprod Bioref.*, 3(6), 601–612, 2009. https://doi.org/10.1002/bbb.182
45. Emil, A., Yaakob, Z., Satheesh Kumar, M.N., Jahim, J.M., Salimon J., Comparative Evaluation of Physicochemical Properties of Jatropha Seed Oil from Malaysia, Indonesia and Thailand. *J Am Oil Chem Soc* 87, 689–695, 2010. https://doi.org/10.1007/s11746-009-1537-6
46. Islam, M.S., Ahmed, A.S., Islam, A., Aziz, S.A., Xian, L.C., Mridha, M., Study on Emission and Performance of Diesel Engine Using Castor Biodiesel. *J. Chemist.*, 2014, 451526, 8, 2014. https://doi.org/10.1155/2014/451526.
47. Ong, H.C., Mahlia, T.M. I., Masjuki, H.H., Honnery, D., Life cycle cost and sensitivity analysis of palm biodiesel production. *Fuel.*, 98, 131-39, 2012. https://doi.org/10.1016 /j.fuel.2012.03.031
48. Abdullah, A.Z., Salamatinia, B., Mootabadi, H., Bhatia S., Current status and policies on biodiesel industry in Malaysia as the world's leading producer of palm oil. *Ener. Polic.*, 37, 5440-48, 2009. https://doi.org/10.1016/j.enpol.2009.08.012
49. Yusuf, N.N.A.N., Kamarudin, S.K., Yaakub Z., Overview on the current trends in biodiesel production. *Ener Conver Manag.*, 52, 2741-51, 2011. https://doi.org/10.1016 /j.encon man.2010.12.004.

50. Raychaudhuri A., Ghosh S.K., Biomass Supply Chain in Asian and European Countries. *Proc. Environ. Sci.* 35 (2016) 914 – 924 . https://doi.org/10.1016/j.proenv. 2016.07.062.
51. Hiloidhari, M., Das, D., Baruah, D.C., Bioenergy potential from crop residue biomass in India. Renew.Sustain. *Ener Revi.*, 32, 504-512. 2014. DOI: 10.1016/j.rser.2014.01.025
52. Ahorsu, R., Medina, F., Constantí, M., Significance and Challenges of Biomass as a Suitable Feedstock for Bioenergy and Biochemical Production: A Review. *Energ.*, 11, 3366, 2018; doi:10.3390/en11123366.
53. Ferreira-Leitao, V., Gottschalk, L.M.F., Ferrara, M.A., Nepomuceno, A.L., Molinari, H.B.C., Bon, E.P.S., Biomass residues in Brazil: Availability and potential uses. *Waste Biomass Valoriz.*, 1, 65–76, 2010.
54. Hamelinck, C.N.,Van Hooijdonk, G., Faaij, A.P.C., Ethanol from lignocellulosic biomass: Techno-economic performance in short-, middle- and long-term. *Biomass Bioener.*, 28, 384–410, 2005. https://doi.org/10.1016/j.biombioe.2004.09.002
55. Lim, J.S., Abdul, Manan, Z., Alwi, S.R.W., Hashim, H., A review on utilisation of biomass from rice industry as a source of renewable energy. Renew. Sustain. Energy Rev., 16, 3084–3094, 2012. https://doi.org/10.1016/j.rser.2012.02.051
56. Ramos, J.L.; Valdivia, M., García-Lorente, F., Segura, A., Benefits and perspectives on the use of biofuels. *Microb. Biotechnol.*, 9, 436–440, 2016. doi: 10.1111/1751-7915.12356.
57. Carneiro, P., Ferreira P., The economic, environmental and strategic value of biomass. *Renew. Ener.*, 44, 17-22, 2012. https://doi.org/10.1016/j.renene.2011.12.020
58. Chen, Y.–H., Chen J.-H., Luo, Y.-M., Complementary biodiesel combination from tung and medium-chain fatty acid oils. *Renew. Ener.*, 44, 305-31, 2012. https://doi.org /10.1016 /j.renene.2012.01.098
59. Faaij, A., Domac, J., Emerging international bio-energy markets and opportunities for socioeconomic development. *Ener. Sustain. Dev.,* 10:7e19, 2006; DOI: 10.1016/S0973-0826(08)60503-7
60. Evans, A., Strezoc, V., Evans, T., Sustainability considerations for electricity generation from biomass. *Renew Sust Energ Rev* 2010;14:1419e27.
61. Field, C.B., Campbell, J.E., Lobell, D.B., Biomass energy: the scale of the potential resource. *Trends Ecol Evol.*, 23(2), 65-72, 2008, doi: 10.1016/j.tree.2007.12.001.
62. Stidham M, Simon-Brown V. Stakeholder perspectives on converting forest biomass to energy in Oregon, USA. *Biomass Bioenerg.,* ;35, 203e13, 2011.
63. Jenssen T. The good, the bad, and the ugly: acceptance and opposition as keys to bioenergy technologies. *J Urban Technol.*, 17, 99e115, 2010.
64. Chung, J.N., Grand challenges in bioenergy and biofuel research: engineering and technology development, environmental impact, and sustainability. *Front. Energy Res.*, 1, 04, 2013. https://doi.org/10.3389/fenrg.2013.00004

65. Virkajarvi, I., Niemela, M. V., Hasanen, A., and Teir, A., Challenges of cellulosic ethanol. *Bioresour.*, 4, 1718–1735,2009.
66. Dale, V. H., Klein, K. L., Perla, D., and Lucier, A., Communicating about bioenergy sustainability. *Energy Manag.*, 51, 279–290, 2013. doi: 10.1007/s00267-012-0014-4.
67. Bátori, V., Ferreira, J. A., Taherzadeh, M.J., Lennartsson, P.R., Ethanol and Protein from Ethanol Plant By-Products Using Edible Fungi Neurospora intermedia and Aspergillus oryzae. BioMed *Resear. Internat.*, 2015, 176371, 10, 2015. https://doi.org/10.1155/20 15/176371.
68. Offei, F., Mensah, M., Thygesen, A., Kemausuor F., Seaweed Bioethanol Production: A Process Selection Review on Hydrolysis and Fermentation. *Fermentat.* 4, 99, 2018; doi:10.3390/fermentation4040099.
69. Harchi, M.E., Kachkach, F.Z.F., Mtili, N.E., Optimization of thermal acid hydrolysis for bioethanol production from Ulva rigida with yeast *Pachysolen tannophilus. South Afric. J Botany.*, 115, 161-169, 2018. https://doi.org/10.1016/j.sajb.2018.01.021.
70. Pereira, S.R., Nogué, V.S.I., Frazão, C.J.R., Serafim, L.S., Gorwa-Grauslund, M.F., Xavier, A.M.R.B., Adaptation of Scheffersomyces stipitis to hardwood spent sulfite liquor by evolutionary engineering Grant Stanley. *Biotechnol. Biofuels*, 8, (1), 50, 2015,
71. John, R.P., Anisha, G.S., Nampoothiri, K.M., Pandey, A., Micro and macroalgal biomass: A renewable source for bioethanol. *Bioresour. Technol.*, 102, (1), 186-193, 2011,
72. El-Mekkawi, S.A., Abdo, S.M., Samhan, F.A. et al. Optimization of some fermentation conditions for bioethanol production from microalgae using response surface method. *Bull Natl Res Cent* 43, 164 (2019). https://doi.org/10.1186/s42269-019-0205-8.
73. Shen, J.C.R., Liao, C., Synergy as design principle for metabolic engineering of 1-propanol production in *Escherichia coli. Metabolic Engineering*, 17. 12-22, 2013. https://doi.org/10.1016/j.ymben.2013.01.008
74. Arshad, M., Hussain, T., Iqbal, M., Abbas, M., Enhanced ethanol production at commercial scale from molasses using high gravity technology by mutant *S. cerevisiae. Braz. J. Microbiol* 48, 403–409, 2017. http://dx.doi.org/10.1016/j.bjm.2017.02.003.
75. Swidah R., Ogunlabi O., Grant C.M., Ashe M.P., n-Butanol production in S. cerevisiae: co-ordinate use of endogenous and exogenous pathways. *Appl Microbiol Biotechnol.* 2018; 102(22): 9857–9866. doi: 10.1007/s00253-018-9305-x.
76. Al-Shorgani, N.K.N., Al-Tabib, A.I., Kadier, A., Mohd Fauzi Zanil, Lee K.M., Kali, M.S., Continuous Butanol Fermentation of Dilute Acid-Pretreated De-oiled Rice Bran by Clostridium acetobutylicum YM1. *Sci Rep.*, 9, 4622 (2019). https://doi.org/10.1038/ s41 598- 019-40840-y.
77. Gallazzi, A., Branska, B., Marinelli, F., Patakova P., Continuous production of n-butanol by Clostridium pasteurianum DSM 525 using suspended and

surface-immobilized cells. J. Biotechnology 216, 29-35, 2015, https://doi.org/10.1016 /j.jbiotec.2015 .10.008.
78. Zhang, H., Wei, W., Zhang, J. Huang, S., Xie, J., Enhancing enzymatic saccharification of sugarcane bagasse by combinatorial pretreatment and Tween 80. Biotechnol Biofuels 11, 309 (2018). https://doi.org/10.1186/s13068-018-1313-7.
79. Silveira, M.H.L., Morais, A.R.C., da Costa Lopes, A. M., Olekszyszen, D.N., Bogel-Łukasik R., Andreaus J., Ramos, L.P., Current Pretreatment Technologies for the Development of Cellulosic Ethanol and Biorefineries. *ChemSusChem.*, 8(20), 3366-3390, 2015. https://doi.org/10.1002/cssc.201500282.
80. Phitsuwan, P., Permsriburasuk, C., Baramee, S., Teeravivattanakit, T., Ratanakhanokchai, K., Structural Analysis of Alkaline Pretreated Rice Straw for Ethanol Production. *Internat. J. Polym. Sci.*, 2017, 4876969, 9 2017. https://doi.org/10.1155/2017/4876969
81. Domanski, J., Borowski, S., Marchut-Mikolajczyk, O., Kubacki, P., Pretreatment of rye straw with aqueous ammonia for conversion to fermentable sugars as a potential substrates in biotechnological processes. *Biomass Bioener.*, 91, 91-97, 2016. https://doi.org/10.1016/j. biombioe.2016.05.008.
82. Shetty, D.J., Kshirsagar, P., Tapadia-Maheshwari, S., Lanjekar V., Sanjay Singh K., Dhakephalkar P.K., Alkali pretreatment at ambient temperature: A promising method to enhance biomethanation of rice straw. *Bioresour. Technol.*, 226, 80-88, 2017. https://doi.org/10.10 16/j.biortech. 2016.12.003
83. Khatri, S., Wu, S., Kizito, S., Zhang, W., Li, J.,Dong, R., Synergistic effect of alkaline pretreatment and Fe dosing on batch anaerobic digestion of maize straw. *Appl. Ener.*, 158, 55-64. 2015. https://doi.org/10.1016/j.apenergy.2015.08.045

3

Engineered Microbial Systems for the Production of Fuels and Industrially Important Chemicals

Sushma Chauhan, Balasubramanian Velramar, Sneha Kumari, Anushri Keshri, Shalini Pandey, Shivam Pandey, Tanushree Baldeo Madavi, Vargobi Mukherjee, Meenakshi Jha and Pamidimarri D. V. N. Sudheer*

Institute of Biotechnology, Amity University Chhattisgarh, Raipur, Chhattisgarh, India

Abstract

The rise in global energy demands and depletion of fossil petro-fuel stocks is making mankind look for alternative green fuels and chemical feedstocks. Although bio-based green technologies are available for the production of biofuels and various commodity chemicals, still, industries are resistant to implement them for production on an industrial scale because of productivity and scale-up cost. The understanding of the molecular infrastructure in the microbial system for the production of fuels and various chemicals provided us new tools for engineering cells for the production of the desired product in engineered non-native hosts. Moreover, productivity could also be enhanced by implementing genetic engineering tools in the native host cells. In this chapter, the efforts are made to provide comprehensive information on the various engineering strategies implemented for the production of biofuels and commodity chemicals in various host systems and their prospective future in scale-up to industrial-scale will be discussed.

***Keywords*:** Biofuels, bioengineering, chemicals, metabolic engineering, synthetic biology

**Corresponding author*: pdvnsudheer@gmail.com; spamidimarri@rpr.amity.edu

Prakash Kumar Sarangi (ed.) Biorefinery Production of Fuels and Platform Chemicals, (59–92) © 2023 Scrivener Publishing LLC

3.1 Introduction

In the modern era, the demand for fuel energy is rising on an exponential scale, however, available fuel resources are depleting around the globe. Moreover, the utilization of nonrenewable fossil fuels like petroleum is adding carbon footprints leading to global climate change. Hence, global nations have come into a common platform in the 2014 Paris Agreement, which was signed for reducing the carbon footprint to at least such an extent that no carbon emissions are added to the globe to maintain the global climate [1, 2]. Under this, all the nations started looking toward green fuels without carbon emissions or with neutral emissions. Bio-based fuels are one among many alternative non-conventional energy resources that can generate fuels without emitting a carbon footprint [3]. Hydrocarbon reserves, besides being used as fuels, serve as feed material for many commodity chemicals with diverse applications. The use of these hydrocarbons in various chemical syntheses is also adding concern to environmental safety. In nature, microorganisms directly and indirectly contribute in the

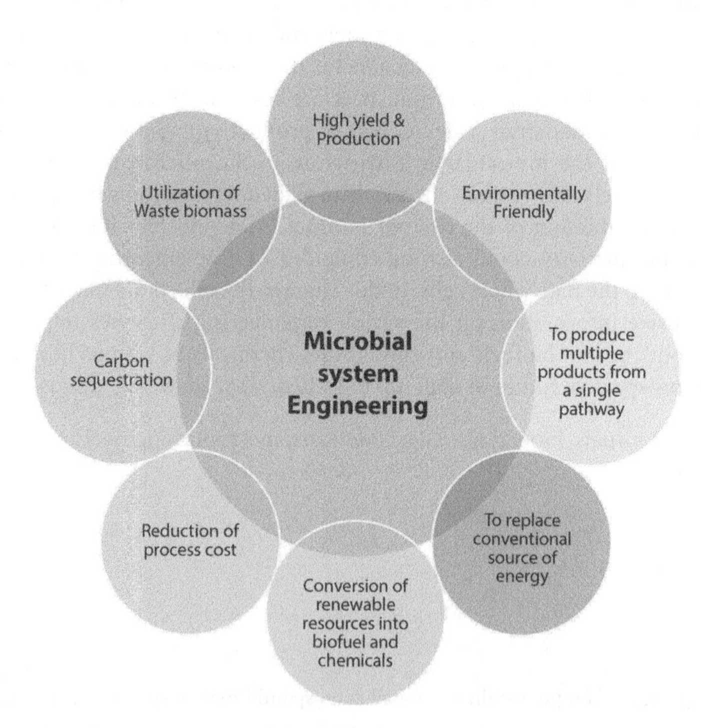

Figure 3.1 Various objectives of microbial system engineering for the production of chemicals and fuels.

production of diverse compounds, such as biofuels and chemicals, and accumulate in the ecosystem naturally. Many microbial cells are reported to produce biofuels such as bioethanol, bio-diesel, bio-gas, bio-hydrogen, and many industrially important chemicals such as amino acids, organic acids, polymers, esters, etc. [3–5]. However, the production rate and yields are not considerable enough to reach commercial demands. Moreover, a major deal with the biological systems to produce biofuels and chemicals is their productivity and dependence on carbon feedstocks for biosynthesis [Figure 3.1].

Traditionally, microorganisms have been improved by selection, random mutagenesis, and screening [6], however, this method has been extremely unpredictable and takes time. In addition to these drawbacks, it is difficult to obtain the desired phenotype by mutagenesis and selection if cell surface biochemistry is not understood properly. Microbial engineering technologies have been spotted as an important research area for producing desired products in considerable yields in natural/non-natural host cell systems. Presently, modern cell engineering technologies allow us to create a novel engineered host to produce the desired product by installing gene clusters synthetically. These microbial engineering systems are originating as a major tool of biotechnology for the green synthesis of chemicals and fuels. Microbial system engineering is gaining crucial pavement in developing sustainable green technology, occupying a significant part in the upcoming green industrial revolution. However, discovering new strategies and including more microbial systems for the production of chemicals and fuels needs to be extended to utilize renewable biomass sources. Autotrophic microorganisms can be used to obtain certain biofuels (bio-hydrogen) and fuel precursors (biomass, starch, lipids, etc.), whereas heterotrophic microorganisms are helpful in obtaining fuels and chemicals from organic matter [7]. Because of low yields, these products could not be commercialized, but literature suggests that metabolic engineering of some microorganisms can lead to improvement in yields which could be commercially profitable.

The gaining of knowledge of cellular mechanisms via modern science and the availability of tools for cell engineering has opened new gates for synthesizing biofuels and chemicals in a carbon-neutral process which could result in sustainable green methods that could compensate for rising demands. With these reliable technologies, the engineered microbial systems can sort the solution for replacing petroleum-based fuels in the near future [8]. Genetic engineering tools can be used for improvising the bio-catalytic abilities of microbial cells as concepts of cell chemistry improvement and this provides microorganisms enhanced ability to increase product titer. Many researchers even tried to introduce genetic elements to the microbial system to establish the production of new products in the engineered cells. Biochemical pathways can be improved by manipulating reproductive and regulatory enzymes, permitting engineered species without any accumulation of undesired mutation. Metabolic engineering together with genetic engineering has become a core method in the production of renewable energy sources such as biofuels and industrially important chemicals in the modern era of biotechnology.

In this chapter, we present the important group of chemicals and fuels produced by engineered microbial fuels. Comprehensive information on how engineering technology has assisted the production of various chemicals and fuels via the green process is presented. This chapter is also included with recent decade developments taking place in product enhancement in the native host systems via engineering and/or introducing new genetic elements to make the engineered system produce novel chemical products like diacids and pigments in our recent studies [9–11]. The chapter also included the bottlenecks and prospects to get the engineered microbial system to reach sustainable industrial implementation to produce biofuels and chemicals for our needs.

3.2 Microbial Systems for Biofuels and Chemicals Production

Since ancient civilization, biofuels have participated in fulfilling our energy needs. Most important among these fuels are alcohols produced by microbial fermentation. Short-chain alcohols such as ethanol and methanol are used as biofuel for the transportation infrastructure. However, they are facing a hard time competing with traditional fossil fuels since we have already established a petro-centric infrastructure and replacing it with biofuels is costly. Nevertheless, one way of overcoming this is by improving productivity and quality. On the other hand, the synthesis of many commodity chemicals is based on petro-derived raw materials and various chemical catalysis reactions. The production industry's reliance on fossil fuel has increased several-fold for the modern industrial process in the last couple of decades. In addition, the depletion of the petro-fuel base across the globe is worrying the chemical industry. Environmental and global warming issues are fueling the industry's worries, facilitating the green synthesis of many accommodative chemicals and raw materials. The additional knowledge of many microbial systems and their ability to produce various chemicals is showing new paths that could be explored to produce many chemicals free of carbon footprint release to the environment. Implementing the microbial system could bypass our reliance on fossil fuel resources and add fuel security via utilization of renewable biomass for the production of chemicals and fuels.

The major hindrance for applying the microbial process on an immediate basis at the industrial level is productivity and downstream processing. This could be bypassed by biosystem improvement by molecular engineering of the host system. The developments in synthetic biology have opened new gates for improvement in productivity and allowed us to create synthetic genetic circuits for producing desired chemicals in non-native hosts [Figure 3.2]. The application of bioengineering of microbial systems has attracted many researchers and industrial stakeholders to develop novel cellular systems for the production of fuels and commodity chemicals. The efforts in engineering the microbial system successfully established various novel engineered microbial systems for the production of various fuels and chemicals and allowed us to improve the productivity several-fold from the titers of μg/L to mg/L. Many chemicals like 3-propanediol, 1–4-butanediol, and fatty alcohols reached a commercial level providing new hope to future generation to replace the non-renewable and environmentally noxious industrial process. These green, carbon footprint-free, environmentally friendly, and renewable biological processes using engineered microbial systems can rule the next generation chemical industry. This will reduce our reliance on fossil fuels and provide energy security, which helps in the recovery of global pollution and turbidity.

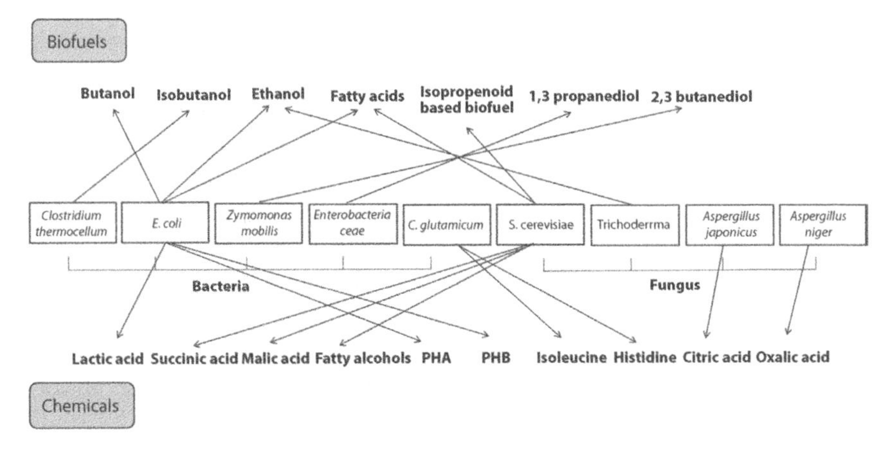

Figure 3.2 Bioengineering of various microbial systems for production of diverse products (biofuels and chemicals).

3.2.1 Microbial Systems for Genetic Engineering and Cellular Fabrication

In the past half a century, the efforts of microbiologists working in the applied field discovered various microbial species with the ability to produce diverse commodity chemicals and fuels and many chemical molecules that act as raw materials for the production of desired target products. To make the process green and sustainable, the researchers made efforts to fabricate the genetic scaffolds in the cellular system for the introduction of a novel pathway for the desired chemical production or enhancing productivity to achieve an industrially scalable and economically feasible system.

Implementing cell engineering in native hosts is always a challenging task. The majority of the efforts seen in this section of microbial engineering are mostly via cellular metabolic pathway fabrication in nonnative hosts, majorly E. coli and other model organisms. This is because of facile methodologies available for the model system via knocking out the genes or overexpression of selected genes of the pathway which are crucial and rate-limiting notches in the synthesis of the desired product. The combination of cell system engineering and synthetic biology made these processes easy and efficient. Nevertheless, the majority of the researchers look to establish the concepts of ideas in the novel product establishment in non-model organisms like Bacillus, Yeast, Aspergillus, and, in some cases, cyanobacterial systems for various chemicals and fuel production. There is a debate of choice between selecting the native strain for product improvement or establishment of the synthetic system by including the alien pathway into a model organism, which cannot be concluded and superiors are too young to conclude since synthetic biology is still in the progress of development. Nevertheless, this book chapter includes both conditions and comprehends all the developments taken in microbial engineering of both native and model systems to produce chemicals and fuels.

The most crucial part of the project is a selection of the biological system for the production of the target product. The right host must be selected to produce a chemical product with desired specifications, which includes functional groups, side chains, chirality, etc. [Figure 3.3]. This selection plays a great role in developing successful technology that could be profitable in the commercial market, e.g., if the application of the target product is for biological purposes, the optical status of the product is very crucial in the biological system. This is very true if the chemical products are in the application of the pharmaceutical industry [12].

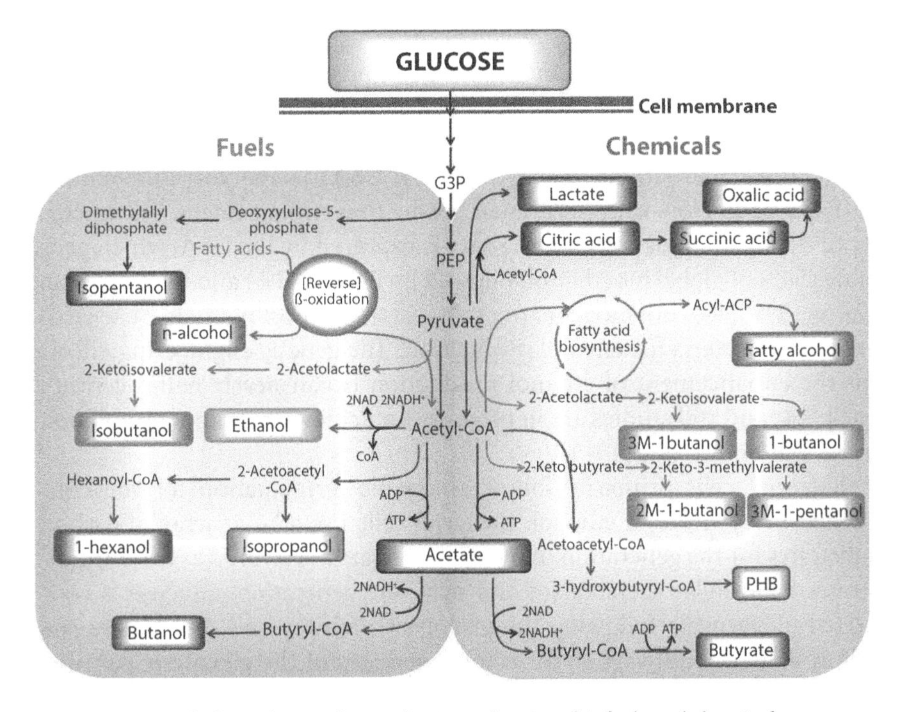

Figure 3.3 Metabolic pathways for production of various biofuels and chemicals.

3.2.2 Engineering of Microbial Cell Systems for Biofuels Production

The major interest of microbial engineering in the modern era of biosystem engineering is centered on the production or enhancement of biofuels and/or their blends. The present context of environmental turbulence and depletion of fossil resources is paramount to the demand for green and renewable fuels. Microbial production of fuels is a potential answer for the creeping demands on fuel energies. Hence efforts were put forward by every nation to establish reliable technology to reach scalable efficiency for the commercial production of biofuels. In this section, various efforts for applying synthetic biology for the production/improvement in productivity of various biofuels are discussed.

3.2.2.1 Alcohols

3.2.2.1a Bioethanol

Bioethanol is the first choice of biofuels that comes into the picture in this contest. From the ancient civilization of humans, alcohols have been used as fuel resources. These can either be used as a direct fuel or as blends for existing Petro-fuels. Ethanol is the most popular and the history of its microbial production has centuries-old technology. It ranks first among the biofuels presently; its global production reached 115 billion liters (according to IEA, International Energy Agency, USA) in the year 2019 and is anticipated to grow with 6% annual production enhancement. The majority of the bioethanol production share is captured by the USA, which generates 56% of global bioethanol followed by Brazil (28%) and the European Union (5%) [13]. Ethanogenic microbial systems most popularly use yeast and some bacteria for ethanol production. The genetic engineering efforts for the enhancement of ethanol production began nearly half a century back [14] and still studies are in progress for achieving ethanol production in various conditions [15–17].

Bacterial cells primarily follow anaerobic fermentation for the conversion of pyruvate to ethanol. Bacterial cells use a large array of carbon substrates for the generation of pyruvate. Sugars are favored as carbon substrates for cellular metabolism and pyruvate generation. Glucose is converted to pyruvate following the glycolysis pathway and different sugars follow axillary pathways to form intermediates of the glycolysis pathway to pyruvate in the cell. The native strains of ethanologenic bacteria, like *Zymomonas mobilis,* produce ethanol in higher yields than *Saccharomyces cerevisiae* since its cellular architecture allows bacteria to sustain a higher percentage of ethanol in the medium. Moreover, *Z. mobilis* has the ability to produce ethanol in aerobic conditions unlike another bacterial systems, as seen in *E. coli,* which needs anaerobic conditions for ethanol production. Though *Z. mobilis* has the ability to accumulate high yields of alcohol, the cellular system is not installed with the utilization of pentose sugars. The significance of pentose sugars in microbial fermentation is important since they can be generated from the lignocellulosic biomass. Zhang and co-workers engineered the *Z. mobilis* by installing pentose utilizing operons under a lac promoter and successfully utilized arabinose and xylose up to 84% along with glucose in mixed sugar fermentation [18]. In addition, upon installation of the overexpression system for expression genes, cellulose and glycosyl hydrolases from *Acidothermus cellulolyticus,* E1, and GH12, respectively, in *Z. mobilis* cell proved to be suitable for cell biomass generation and ethanol production [19]. Unlike *Z. mobilis,* the model organism *E. coli* has an inherent pathway for the production of ethanol utilizing pyruvate generated from pentose sugars. The advantage of the *E. coli* host is it can utilize a large spectrum of carbon sugars for cellular metabolism. The inherent pathway of *E. coli* for the production of ethanol is inefficient for accumulating higher levels in native conditions since a competitive pathway of lactate accumulation dominates. The first effort in the engineering of *E. coli* was conducted by Ingram in 1987, where knocking out the genes *pdc* (pyruvate decarboxylase) and *adhB* (alcohol dehydrogenase) allowed *E. coli* to accumulate ethanol. However, the limited regeneration of NADH caused a bottleneck in the *E. coli* [2]. Kim *et al.* (2007) developed a mutant strain of *E. coli*, K-12 strain, which could accumulate ethanol via homoethanologenenic fermentation under glucose and xylose utilization and the mutations are mapped and found to be in pdh operon (*pdhR aceEF lpd*), which codes for the pyruvate dehydrogenase complex [20]. However, this pathway is possible in anaerobic conditions, which limits its application in industrial scale up. Hence, researchers are looking towards engineering cells to accumulate ethanol in aerobic conditions. In this context, the notable work was put forward by Fithriani *et al.*, [17] where the group made this possible by expressing pyruvate decarboxylase and alcohol dehydrogenase from *Z. mobilis.*

Besides bacterial species for ethanol production, eukaryotic systems are also well explored for ethanol production. The most popular one in this context is S. cerevisiae. However, the cellular infrastructure did not allow it to utilize pentose sugars. Hence, the important efforts for building the cellular ability to utilize the pentose sugars were conducted in *S. cerevisiae.* Considerable efforts are put forward to include the genetic elements to utilize pentose sugars. Introducing the xylose isomerase enables *S. cerevisiae* to utilize the xylose to produce ethanol [21, 22]. In addition to the addition of genes to metabolism circuits, alcohol utilizing genes were targeted for enhancing the final titer in the medium. Alcohol dehydrogenase Adh2p encoded by the ADH2 gene catalyzes the ethanol to aldehyde. Hence, many researchers targeted the Adh2 gene or its regulator gene for making engineered cells devoid of reaction conversion from alcohol to aldehyde [23]. Many approaches were explored for the inactivation of ADH2 in *S. cerevisiae.* Among these, PCR-mediated homologous recombination is one of the successful methods in inactivating ADH2. However, the efficiency and time consumption of the gene manipulation is tedious and advancements are needed for high throughput application for genetic engineering. Modern techniques like TALEN (Transcription activator-like effector nuclease) CRISPR-Cas9 based gene-editing technologies are making the process more facile and efficient and are opening new gates for genetic improvement of the target strains [24]. The lack of genetic tools to utilize lignocellulosic sugars in *S. cerevisiae* made scientific society look for other fungal sources for ethanol production. The focus on other fungal species with the ability of ethanol production utilizing lignocellulosic sugars (pentose sugars) are ever in demand and species like Fusarium and Neurospora are suitable for ethanol production from pentose sugars. *F. oxysporium* is the most explored species for ethanol production, however, species like *F. verticillioides, F. equiseti*, and *F. acuminatum* could also accumulate ethanol. There are many studies in fermentation strategies for enhancement of the yields, however, the genetic fabrication trails in these species were very scarce because of limited tools available to apply to these species.

Though the efforts of genetic engineering resulted in achieving ethanol yields near to the theoretical values or from waste biomass like lignocellulosic sugars, the major bottlenecks with ethanol fuel are its low energy (70%) content compared to gasoline, high vapor pressure, and non-compatibility to the present existing engines, making it not a readily replaceable option for petrol-based fuel engines. In addition, its natural property of absorbing moisture results in corrosion of the engine. Moreover, with the present technology, the distillation process is cost-intensive and not economically competitive, hence in this area more innovation is needed to make biofuel commercially viable. Hence, the focus shifted towards the production of advanced alcohol biofuels like higher alcohol production using microbial systems. These higher alcohols, also termed advanced biofuels, are next-generation biofuels since these can readily be utilized in present running transportation engines with no or slight modifications.

3.2.2.1b Higher Alcohols

The higher alcohols are also called advanced biofuels because of their compatibility with existing engine designs which are based on petro-fuels. Moreover, these could also be transported with a similar distribution chain. This will save much of the infrastructural investments for their commercialization. In these higher alcohols, unlike ethanol, the cells depend on a fatty acid or amino acid metabolism. In general, fatty acid metabolism is used to produce linear alcohols while amino acid metabolism leads to branched-chain alcohols. Among these higher alcohols, 1-butanol occupies a significant interest in commercialization because of its higher-octane number and its energy content near to gasoline makes it a potential replacement for commercial petro-fuel. Taking the case of butanol, its high combustion value is equivalent to 84% of the gasoline combustion value and is completely miscible with gasoline. Blending this alcohol has proven to have a significant effect on reducing emissions from the combustion engine. Moreover, 1-butanol is naturally produced in many bacterial systems, making this a primary target for the genetic engineering of cells for the betterment of productivity or production utilizing waste biomass.

Clostridium is one of the most popular species which produce 1-butanol by utilizing the ABE (acetone-butanol-ethanol or cold channel) pathway. The simple sugars are utilized in this pathway to produce acetyl-CoA and two molecules of these are converted to butyryl-CoA. This will be converted to 1-butanol with the help of crucial enzyme butyraldehyde/butanol dehydrogenase, a dual functional enzyme [25, 26]. Lack of genetic tools for the metabolic engineering of *Clostridium* species hindered the strain improvement via genetic engineering. Many of the advancements seen in the last decade focused majorly on enhancing productivity via reducing the byproducts of the pathway like acetate, butyrate, and acetone. However, later it was found that increasing the flux towards 1-butanol indeed did not fetch a higher accumulation of target, rather the flux in the cells was diverted to acids [27–29]. *C. acetobutylicum* also produces 1-butanol via a hot channel pathway where acetyl-CoA is converted to butyryl-CoA then to butanol in a process called solventogenesis. The hot channel of butanol production can be enhanced by simple disruption of two gene *pta* and *buk* codes for phosphotransacetylase and butyrate kinase. This has increased nearly 9-fold in the titer of 1-butanol via hot channel flux. Overexpression of mutated aldehyde/alcohol dehydrogenase resulted in reinforcing the direct butanol-forming flux and resulted in enhancement of the titer up to 245% compared to wild type. Upon fermentation under glucose via batch fermentation, the final productivity was found to be 18.9 g/liter [30]. Alternatively, Hou *et al.* tried blocking the acid by-product in the cold channel pathway by blocking *adc* genes (acetoacetate decarboxylase) responsible for acetone formation and this resulted in the reduction of acetone production from 2.64±0.22 g/L to 0.15±0.08 g/L. In addition, expressing the glutathione-encoding genes (*gshAB*) from *E. coli* resulted 1-butanol from 5.17±17 g/L to 8.27±0.27 g/L [31]. Even simple over-expression of the entire pathway of 1-butanol synthesis resulted in 14.9 g/L without any gene deletions [31]. Using mannitol reduced simple sugar and increased the 1-butanol titer in *C. tyrobutyricum*. Overexpression of the *adhE2* gene from *C. acetobutylicum* in *C. tyrobutyricum* without any gene deletions resulted in titer as high as 21 g/L.

The lack of suitable genetic tools for the natural host's gene manipulation made strain improvement too slow. Hence, researchers made efforts towards installing the genetic infrastructure in the model microbial system with non-native hosts like *E. coli* and other organisms for 1-butanol production. However, the results are not encouraging probably due to the fact that the activity of the *Clostridial* genes in non-native hosts is not sufficiently active. However, various studies in engineered *E. coli* by expressing *adhE2* (bifunctional aldehyde/alcohol dehydrogenase), *crt* (3-hydroxybutyryl–CoA dehydratase), and *hdd* (3-hydroxybutyryl [HB]–CoA dehydrogenase) from *C. acetobutylicum* in anerobic fermentation, along with formate dehydrogenase resulted in accumulation of 1-butanol up to 15g/L in batch fermentation. The major hindrance and limiting factor in the cell for the production of 1-butanol is the limited availability of NADH. Some organisms are capable of producing sufficiently enough NADPH (e.g., cynobacteria could generate via photosynthesis) by including an NADPH-dependent enzyme cluster that will have the ability to generate 1-butanol. Hence, studies are also extended to include the *Hbd* gene derived from *Clostridia,* which is an NADH-dependent enzyme along with aldehyde and alcohol dehydrogenases in Cyanobacteria, which resulted in enhanced 1-butanol from CO_2. This will add the advantage of carbon capture for the production of biofuel.

Next to the butanol, isopropanol occupies a significant segment in fuel blending. Because of its lower hygroscopic nature and higher energy density of 23.9 MJ/L relatively higher than ethanol, it is more preferable in blending and in some instances can replace gasoline. Hence, the search for the isopropanol producing bacteria was made and several species of *Clostridium* are reported to produce isopropanol. Indeed, 52 different strains of *C. beijerinckii* are listed in the literature with the ability to produce isopropanol [32]. However, the native strains resulted in very limited production titer as high as 30mM. Limited knowledge on the metabolic regulation of the isopropanol synthesis hindered initial efforts in yield improvements, however, some efforts are paid off with strain improvement in physiological tolerance and productivity towards isopropanol. In native strains like *Clostridium,* the isopropanol is generated using the ABE pathway. The engineering studies in native strains of *Clostridia* species in the last decade denote advancement in molecular tools. In *Clostridium,* acetone is accumulated as a byproduct during ABE (Acetone-Butanol-Ethanol) fermentation. The isopropanol production is by alcohol dehydrogenase activity taking acetone as substrate mostly conducted by NADPH-dependent primary/secondary alcohol dehydrogenase. Hence, the fraction of propanol percentage in total alcohols accumulated during fermentation is very limited, not more than 20 percent of the total alcohols. There are some species, like *Propinibacteria,* that can accumulate selectively propionic acid as a major fermentative end-product via implementing selective conditions along with the n-propanol as a co-product. In *Clostridia* species, the major selective product in the alcohol is butanol and acetone as a by-product. By introducing the enzyme, dehydrogenase activity taking acetate as substrate can produce isopropanol upon fermentation. There are multiple studies conducted utilizing this hypothesis by overexpressing primary/secondary alcohol dehydrogenase from *C. beijerinckii* NRRL B593, acetoacetyl-CoA: acetate CoA-transferase and acetoacetate decarboxylase enzymes resulted in higher titer of isopropanol along with butanol and ethanol. The engineered strain accumulated nearly 8.8 g/L of isopropanol from 90 g/L glucose in anaerobic fermentation [33].

The researchers also tried to produce isopropanol in non-native producers and most of the studies are conducted to understand the process in model organisms rather than high productivity. These studies did not significantly surpass the efficiencies of native strains, however, they added knowledge and understanding of the pathways and regulation of the enzyme complexes involved in the product accumulation [33]. Two independent studies conducted in *E. coli* were reported. This was made possible by extending the clostridial acetone reduction pathway. Compared to the SADH (secondary alcohol dehydrogenase) from *Thermoanaerobacter brockii, C. beijerinckii* alcohol dehydrogenase gave higher isopropanol yields [34, 35]. The productivity of these recombinant strains are further increased by removing the product by continuous gas stripping and yields are achieved up to 143 g/L in fed-batch fermentation [36]. Implementation of flux repartitioning in the cell via cooperating with the TCA and isopropanol pathway by genetic toggle switch resulted in a 3-fold increase in isopropanol accumulation compared to the control [33, 37].

Yeast cells are native alcohol producers and have naturally developed a stress tolerance system towards alcohols. Hence, the researchers also tried to install genetic tools to produce isopropanol in yeast and candida species. The recent studies by Tamakawa *et al.* introduced that isopropanol pathway genes from *Clostridia* (*adc*-acetoacetate decarboxylase, *cftAB*- acetoacetyl-CoA:acetate CoA-transferase, and the *adhB593*-acetone-accepting alcohol dehydrogenase) resulted in the accumulation of isopropanol, however, the titers are very low. Further, the productivity is increased several-fold by simple way up on expression acetyl-CoA synthetase and acetyl-CoA transferase which improved acetyl-CoA production in cytosol, resulting in reaching titer up to 27.2 g/L isopropanol on fed-batch fermentation taking 200 g/L glucose as substrate. Nevertheless, the productivity was lower than *E. coli*, however, the window of improvement is still expected if the mitochondrial activities are reduced by implementing anaerobic conditions in yeast [33, 38].

3.2.2.1c C5-C10 Alcohols

The lower carbon alcohols are very popular and prove to be useful as biofuels. Still, they are limited to a degree of blending fuels rather than competing with petro-fuels and replacing them. This is because of their high hygroscopic nature and less energy content than gasoline. Unlike lower carbon alcohols, higher alcohols like C5 or above show explicitly low/no hygroscopic nature and holding energy densities near to the gasoline makes them more compatible with the fuel infrastructure presently established to the petro-fuel base. However, the major bottleneck is that in nature, the biological sources have limited production genetic infrastructure in the cellular cradle. There are few pathways availed for their production and in this section, we will put forward a brief discussion on cellular pathways responsible for generating these higher alcohols. We will also discuss efforts of the various researchers towards genetic engineering of cells for production and yield enhancement of these higher alcohols.

Bacteria like *Clostridium* species utilize the β-oxidation cycle to breakdown (Cn+2)-acyl-CoA to acetyl-CoA and (Cn)-acetyl-CoA by reversal of the β-oxidation cycle higher-chain alcohols and other chemicals. Dellomonaco *et al.* [39] demonstrated this for the production of higher alcohols. In their study, the authors used endogenous dehydrogenases and thioesterases to synthesize n-alcohols along with fatty acids and 3-hydroxy, keto, and trans carboxylic acids. In addition to this strategy, the higher alcohols are also demonstrated to be able to produce via fatty acid biosynthetic pathways. To implement this, various strategies were designed and successfully demonstrated. Fatty acids are synthesized from the thioesters and upon converting these fatty acids in fatty acyl-CoAs, will further be converted to higher alcohols by introducing acyl-CoA reductase and aldehyde reductase [40, 41].

In a further step, Zhang *et al.* (2008) explored the 2-keto acid pathway for the production of higher-chain alcohols which are above 5 carbons. This is made possible by expanding the ability of substrate chain length acceptance of 2-isopropylmalate synthase (LeuA) and 2KIV decarboxylase derived from *Lactobacillus lactis* using protein engineering tools [42]. It is the first study to show the production of non-natural alcohols produced in the engineered microbial system, which demonstrates the strength of synthetic biology advancement in modern science for creating non-natural fuel commodities in the biological green process. This study successfully produced 3-methyl-1-pentanol up to the yields of 793.5 mg/L. Further, including a few more mutants created by rational and semi-rational protein engineering methods allowed it to produce 4-methyl-1-hexanol up to the titer 51.9 mg/L.

3.2.3 Engineering of Microbial Cell Systems for Chemical Synthesis

The commodity chemical synthesis as old as urea synthesis was developed by Wohler in 1928. In general, the synthesis of various commodity chemicals involves extreme physical conditions and the presence of suitable catalysts. In a majority of conditions, the process is not environmentally friendly. Moreover, the raw materials are significantly derived from petroleum derivatives which is again a matter of concern because much of the carbon emissions globally are because of petroleum processing and usage. Biosynthesis of many chemicals has been encouraged for the past couple of decades and researchers identified different microbial sources which can be utilized for the synthesis of a great array of chemical commodities. Nevertheless, the productivity and titers of the accumulated chemical products are less economically non-competitive on a commercial scale. Progress in the developments of metabolic engineering, synthetic biology, and cell fabrication technology have given the hope of making biological process commercially viable via enhancing productivity and utilizing renewable biomass of feedstocks. Moreover, government policies are making attractive investments for this green synthesis process via funding infrastructure for establishing biological processes and encouraging engineering of cells for the synthesis of commodity chemicals. The major focus in engineering biological systems is to (i) enhance the productivity and (ii) utilization of waste biomass for whole-cell growth and chemical synthesis. Among the important chemicals, organic acids come as the first segment of chemicals which are tried from long before modern biology evolved and followed by many more like alkenes, hydrocarbons, fatty alcohols, isoprenoids, etc. In this section of the chapter, discussions were made on how the microbial systems are engineered for product enhancement in the native system or engineering non-native hosts for the installation of genetic segments for desired product synthesis.

3.2.3.1 Organic Acids

Organic acids are the most common commodity chemicals used in both domestic applications and as industrial raw materials or building blocks. A great list of organic acids such as acetic acid, formic acid, succinic acid, lactic acid, oxalic acid itaconic acid, citric acid, etc. are produced from microbial sources from many native species. However, the overall commercial level of production is occupying only to 5-10% on an industrial production scale and a potential window is availed for enhancing the productivity via genetic engineering [5, 43]. In this contest, the most prominent species for engineering is *E. coli* for the production of various organic acids. *E. coli* being a facultative anaerobe has the inherent genetic infrastructure for the production of some organic acids if the external electron acceptor oxygen is absent. In general, *E. coli* uses simple sugars for generating ATP and NADH molecules via glycolysis and, in the presence of oxygen, pyruvate upon glycolysis is converted to acetyl-CoA and CO_2 by pyruvate dehydrogenase complex and proceeds to the citric acid cycle. However, in absence of oxygen, this complex enzyme is down-regulated and uses various substances as electron acceptors. In this condition, pyruvate formate lyase will take over and catalyzes the formation of formate and acetyl-CoA and under fermentative conditions. *E. coli* can accumulate organic acids like formate, acetate, and lactate, along with some amount of ethanol to maintain redox balance in the cell [1]. However, an *E. coli* mixture of organic acids are accumulated upon anaerobic fermentation and genetic engineering strategies enabled the strain to achieve single acid accumulation. Causey *et al.* (2003) successfully developed the *E. coli* of homoacetate accumulation capability. In this study, the PFL complex, fumarate reductase, and lactate dehydrogenase were disabled by deleting the genes *focA-pflB*, *frdBC*, and *ldhA* and *adhE* (alcohol dehydrogenase gene) eliminated the competing reaction of acetyl-CoA. The citric acid cycle was blocked by removing the gene codes for α-ketoglutarate dehydrogenase and resulted in the accumulation of 86% acetate selectively taking glucose as a carbon source [44].

Besides acetate, lactic acid is also of biotechnological interest due to its involvement as a precursor for the synthesis of PLA (polylactic acid), a polymer popularly termed as green plastic. The biological synthesized lactic acid is preferred for PLA synthesis since the properties of the PLA vary with the optical isomer used for the synthesis. Lactic acid can be synthesized chemically, however the product will be of a mixture of both optical isomers. Hence, a biologically synthesized one is always preferred for PLA synthesis. The selective fermentative synthesis of lactate was achieved by blocking the competing pathways by deleting gene codes for phosphotransacetylase (*pta*) and phosphoenolpyruvate carbozylase (*ppc*). This double mutant strain derived from *E. coli* RR1 could be able to accumulate 90% yields, taking glucose as a carbon source [45]. Blocking of lactate dehydrogenase in the native strain and introduction of L-lactate dehydrogenase from *Lactobacillus casei* in *E. coli* resulted in the production of pure L-lactate, proving that either of the lactate could be possible to produce selectively using *E. coli* as a host. The majority of these studies used glucose as the carbon source, however, to make the product economically feasible at the industrial level, utilization of waste biomass would be a good platform for industrial success. With this view, multiple studies were done to utilize pentose sugars (lignocellulosic sugars) and glycerol (waste byproduct originated during biodiesel synthesis) for the production of lactate. Dien *et al.*, in their study, utilized the mutated B-strain of *E. coli* (Δ*ldhA*, Δ*pta*) with plasmid expression construct carrying L-lactatedehydrogenase from *Streptococcus bovis* could produce 93% conversion, taking glucose and xylose sugars as carbon sources [46]. Introducing the glycerol consuming pathway (glycerolkinase *glpK*, and glycerol- 3-phosphatedehydrogenase *glpD*), using plasmid expression construct, and disabling the lactate utilizing gene product (Δ*lldD*) resulted in producing 0.873 g L-lactic acid per gram of glycerol utilization and is the equivalence of 90% theoretical maximum [47]. Producing succinate in *E. coli* is challenging since it is a citric acid cycle intermediate and the metabolic flux moves towards fumarate as an electron acceptor, hence wild type accumulate is presented in much smaller quantities of succinic acid (0.15 g/g glucose) [48]. Many genetic engineering studies are conducted in this regard, however, the productivity is limited to 0.66 g succinate/g glucose. In addition, poor growth rate and inconsistent yields were observed with *E. coli* as a host [49].

Though many engineered strains based on *E. coli* hosts have been developed, the restrictions in the use of these products generated from engineered *E. coli* hosts restricts use in the food and medical industries because of its endotoxin-producing ability. Alternatively, *Bacillus subtilis* is a good alternative in this context since it does not produce any endotoxins, making it a superior host for genetic engineering to produce organic acids [50, 51]. *B. subtilis* CHI strain with the inactivated *alsS* gene could be able to reach productivity up to 1.05 g of lactic acid with one gram of glucose in batch fermentation. One important organic acid produced by *B. subtilis* is KG-A (α-ketoglutaric acid), which is a significant chemical building block in the synthesis of many industrially important chemicals. Hossain *et al.* demonstrated whole-cell biocatalysis to produce KG-A from feed substrate L-glutamic acid by overexpressing the protein L-amino acid deaminase (AAD), resulting in productivity up to 4.8 g/L [52]. By using an engineered AAD enzyme with enhanced activity and disabling the KG-A utilization pathway by mutating SucA enzyme, the *B. subtilis* host was able to produce the KG-A 12.21 g/L from 15g/L glutamic acid [42]. Another important organic acid derived from amino acids is GABA (γ-aminobutyric acid). GABA is a non-proteogenic amino acid that acts as a neurotransmission inhibitor that has various clinical applications in the treatment of neuropathological disorders. It is synthesized by the decarboxylation of L-glutamate by the action of glutamate decarboxylase (GAD). By the heterologous expression of the GAD enzyme derived from *Streptococcus salivarius* subsp. thermophilus Y2, *B. subtilis* is able to produce 5.26 g/L GABA which is of food-grade quality [53].

3.2.3.2 Fatty Alcohols

The selective type of alcohols with hydroxyl groups (-OH) along with ethylene groups or fatty acid derived chains are called fatty alcohols or oleochemicals. These are primarily synthesized via chemical synthesis and, to some extent, they are prepared naturally. These fatty acids contain the acyl chains with hydroxyl moieties which provide these chemicals with various selective properties like amphiphilic nature, etc. These are used in detergents, cosmetics, emollients, creams, soaps, and various personal care product preparation [4]. The first fatty alcohol (cetyl alcohol) was discovered in sperm whale oil in 1817 by Chevreul and successfully separated from the oil component [54]. Later on, many variants of fatty alcohols were reported from plants and algae bacterial species. However, the majority of commercial level productions are mostly utilizing hydrocarbon feedstock via chemical synthesis. Many commercial processes have been developed to synthesize these alcohols from fatty acids derived from vegetable oils utilizing chemical catalysis. The naturally extracted fatty alcohols from the biological route are always preferred and considered to be environmentally safe and green products. However, in natural hosts, these are synthesized in trace amounts and not commercially viable. Hence, many researchers tried to adopt genetic engineering tools to establish an engineered microbial system for the production of various fatty alcohols. In this section, we will be describing a few to elucidate how the methods are adopted for the synthesis of fatty alcohols.

Fatty acid biosynthesis follows a conserved pathway which is most often the same in prokaryotes, eukaryotes, and higher organisms. Besides various compartmentalization, the enzymes evolved are different from organism to organism, however, it all begins with the carboxylation of acetyl-CoA to malonyl-CoA catalysis. The fatty acid synthase complex (FAS) takes malonyl-CoA as an extender and synthesizes fatty acid using acyl carrier protein (Acp). Generally, acetyl-CoA acts as the starting unit, however, in many cases, propionyl-CoA or other branched products derived from branched amino acids can also participate, leading to the odd or branched variants of fatty acids respectively [55]. These fatty acids, once synthesized, can participate in the synthesis of fatty alcohol production in the natural hosts. This biological generation of fatty alcohols takes place via the reduction of fatty acyl CoA or fatty acyl-ACP catalyzed by fatty acyl reductase (FAR) [56]. Alternatively, the reduction of fatty acids by carboxylic acid reductase (CAR) also generates fatty alcohols [57].

As stated earlier, the production levels of these alcohols are in very limited amounts in natural hosts and the rewriting of the metabolic routes promised very much in enhancing the productivity into commercially feasible levels by engineered microbial systems. A few of these studies are detailed here. The most successful studies are in the model system using *E. coli*. *E. coli* does not produce fatty alcohols naturally, but the required enzymes derived from other sources has successfully produced fatty alcohols.

3.2.3.3 Bioplastic

Since the 1970s, there has been an exponential growth in the plastic industry and in starting of the past decade the production rate has doubled every year due to the global demand. However, soon the detrimental effects of synthesized plastic due have caused accumulation on an ecological level because of non-degradability. Bioplastics like polyhydroxyalkanoates (PHAs) and polyhydroxybutyrate (PBHs) are identified as alternatives for this synthetic plastic. Although these have been known for a long time, polyhydroxyalkanoates have been produced by different bacteria and their commercial efficacy remains challenging due to unstable thermo-mechanical properties and high-cost production issues [58, 59]. Hence, metabolic engineering of the microbial system as a tool was explored to enhance productivity and make it a commercially viable product in the market.

As a proof of concept, multiple studies were made in various organisms including *E. coli*. A slow conversion rate of the substrate to polyhydroxyalkanoates is the major cause of high-cost production which can be overcome by the engineering of microbes. 3-Hydroxybutyrate (3HB) containing polyhydroxyalkanoates requires fatty acid as a substrate for short and medium chain length formation of 3-Hydroxybutyrate. However, in the cell, most of the fatty acids are converted to acetyl-CoA for other metabolic processes. This hampers the production rate of polyhydroxyalkanoates [60]. This can be resolved by deletion of genes associated with a β-oxidation pathway in the cell. The selected genes *fadA* and *fadB* were tried to block the β-oxidation pathway in *Pseudomonas putida* and *P. entomophila* led to a higher accumulation of 3-hydroxyacyl-CoA intern, which increased the accumulation of polyhydroxyalkanoates [61]. The accumulation of p(3HB) in non-native hosts like *E. coli* is also tried by expressing the gene clusters containing succinate degradation of *Clostridium kluyveri* and P(3HB) accumulation pathways of *Ralstonia eutropha*. The resulting recombinant *E. coli* produced P(3HB), taking glucose as carbon substrate. In addition, the deletion of the native succinate semialdehyde dehydrogenase genes (*sad* & *gabD*) increased C-flux for 4-HB production and demonstrated that the tetramer of 3-HB-co-3HV-co-4HB can be obtained directly from lactose or waste raw material.

Besides the improvement of the product, genetic engineering of the strains was also conducted for improvement in the thermomechanical properties of the PHA monomer. Engineered *P. putida* KT2442 (with deleted β-oxidation) synthesized a stable monomer ratio and molecular weight PHA with 3-PHB and 3HHx monomers. Similarly, engineered *P. entomophila* LAC23 when grown on different fatty acid precursors synthesized stable monomers of C5 (3- hydroxyvalerate) to C14 (4- hydroxytetradecanoate) [62]. Though the stories for microbial engineering in context to polymer synthesis are plenty, many more strategies need to be explored for making it commercially viable. This could be achieved by installing the methods utilizing waste biomass for product accumulation rather than using valuable sugars for cell growth and monomers synthesis.

3.3 Conclusions

With the extensive use of fossil fuels from the industrial revolution, the global carbon footprint has accumulated to a breaking point. The further reluctance to shift from these petro-centric fuels to green fuels may cause irreversible damage to the globe. With the current acceleration of development and the technologies in the biological field, the proof of concept to produce fuels and chemicals was availed (Table 3.1). However, now to implement on an industrial scale, the limiting factor is developing infrastructure. Hence, governments need to be investing their country's GDP for the establishment of the green process for fuel and chemical production. Many developing countries are investing in this and contributing their part in reducing the carbon footprint on the globe. However, the developing countries are still not at a point to fund the infrastructure to implement bioengineering technology in the industrial field. Hence, the policies should be designed in such a way that the beneficiaries will be on a global level. In addition, more innovations need to be delivered in the reactor engineering sector and industrial scaleup to make this green product get into the market.

Table 3.1 List of some engineered microorganisms and productivities of fuels and chemicals.

S. no	Biofuels	Metabolic pathways	Microorganism	Type of use	Strategy	Yield	Reference
1	Methyl-1-pentanol	Melvonate Pathway	*E .coli*	Biodiesel	Expression of pyrophosphatase nudB gene isolated from B. subtilis		[42]
2	1-butanol	ABE Pathway	Clostridium		Overexpression of mutated aldehyde/alcohol dehydrogenase & blocking *adc* gene	2.64±0.22 g/L to 0.15±0.08 g/L	[25, 26]
3	Fatty Alcohols	Fatty Acid Biosynthesis	*E. coli*	Diesel substitute	Overexpression of fadD and an acyl-CoA reductase from Acinetobacter Calcoaceticus (acr1) and reduction of fatty acids by carboxylic acid reductase (CAR)		[56]

(Continued)

Table 3.1 List of some engineered microorganisms and productivities of fuels and chemicals. (*Continued*)

S. no	Biofuels	Metabolic pathways	Microorganism	Type of use	Strategy	Yield	Reference
4	Fatty Acid Ethyl Esters (FAEEs)	Fatty Acid Biosynthesis	*E. coli*	Diesel substitute	Overexpression of endogenous *fadD* gene together with a wax-ester synthase gene *(atfA)*	427 mg/L	[56]
5	Isopropanol	CoA-dependent-B-oxidation	*E. coli*	Gasoline	Expression of acetyl-CoA acetyltransferase (thl), acetoacetyl-CoA transferase (atoAD), acetoacetate decarboxylase (adc) and alcohol Dehydrogenase (adh)	27.2 g/L	[33, 38]

(*Continued*)

Table 3.1 List of some engineered microorganisms and productivities of fuels and chemicals. (*Continued*)

S. no	Biofuels	Metabolic pathways	Microorganism	Type of use	Strategy	Yield	Reference
6	1-butanol	CoA-dependent-B-oxidation	Clostridium acetobutylicum	Gasoline	Expression of NADH-dependent enzyme, trans-enyol-CoA Reductase (Ter)	15 g/L	[25]
7	Ethanol	EMP Pathway	*E. coli*	Fuel	knocking out the competing lactate dehydrogenase and expressing a bifunctional acetylaldehyde–alcohol dehydrogenase from *C. thermocellum*		[18]
8	Ethanol	Ethanol Pathway	*E. coli*	Fuel	Expression of pdc and adhB genes of Zymomonas mobilis		[20]

(*Continued*)

Table 3.1 List of some engineered microorganisms and productivities of fuels and chemicals. (*Continued*)

S. no	Biofuels	Metabolic pathways	Microorganism	Type of use	Strategy	Yield	Reference
9	Ethanol	Ethanol Pathway	*E. coli*	Fuel	lactate-producing isolate of KO11, the SZ110 strain, was reengineered to delete all fermentative routes for NADH and insert complete ethanol producing pathway genes pdc, adhA, and adhB into chromosomes	46 g/L	[17]
10	Isobutanol	Keto Acid Pathway	*E. coli*	Gasoline	kivD and adh2 were introduced from *lactobacillus lacti* and *S. cerevisiae*, respectively		[26]

(*Continued*)

Table 3.1 List of some engineered microorganisms and productivities of fuels and chemicals. (*Continued*)

S. no	Biofuels	Metabolic pathways	Microorganism	Type of use	Strategy	Yield	Reference
11	n-butanol	Butanol Synthesis Pathway	*E.coli*		1. Developed butyrate conversion strain by removing undesirable genes, recruiting endogenous *atoDA* and *Clostridium adhE2*	5.5g/L	[4]
12					2. Conversion of acetyl-CoA into butanol by enhancing the acetyl-CoA production and deleting genes whose expressions are needed to convert acetyl-CoA into ethanol and acetate	6.1 g/L	[4]

(*Continued*)

Table 3.1 List of some engineered microorganisms and productivities of fuels and chemicals. (*Continued*)

S. no	Biofuels	Metabolic pathways	Microorganism	Type of use	Strategy	Yield	Reference
13	Isoproponol	L-theronine Pathway	*E. coli*	Fuel	By the conversion of L-threonine into 2-ketobutyrate using *lvA*, *tdc* then 2-ketobutyrate is converted into 1-propanol using *kdc* and *adh*		[34]
14					by deleting competing pathways, stress response genes, and releasing feedback inhibition of amino acid biosynthesis	10.8 g/L	[35]

(*Continued*)

Table 3.1 List of some engineered microorganisms and productivities of fuels and chemicals. (*Continued*)

S. no	Biofuels	Metabolic pathways	Microorganism	Type of use	Strategy	Yield	Reference
	CHEMICALS						
15	Lactic Acid	EMP Pathway	*E. coli*	Chemical building block	By introducing the glycerol consuming pathway	0.873 g/g glycerolk	[47]
16	Acetic Acid	PFL Pathway	*E. coli*	Chemical building block	PFL complex, fumarate reductase and lactate dehydrogenase were disabled by deleting the genes *focA-pflB*, *frdBC*, *ldhA*; and *adhE*	86% acetate in total acids	[44]
17	Citric Acid	EMP pathway	*E. coli*	Chemical building block	Pyruvate formate lyase complex expression		[1]
18	Succinic Acid	Pentose phosphate Pathway	*E. coli*	Chemical building block	Enhancement of the metabolic flux enhancement towards Fumarate	0.15 g/g glucose	[47]

(*Continued*)

Table 3.1 List of some engineered microorganisms and productivities of fuels and chemicals. (*Continued*)

S. no	Biofuels	Metabolic pathways	Microorganism	Type of use	Strategy	Yield	Reference
19	Glutamic Acid	α-ketoglutarate Pathway	*B. subtilis CHI*	Pharmaceutical	Overexpressing the protein L-amino acid deaminase (AAD)	4.8 g/L	[48]
20	α-ketoglutaric Acid	α-ketoglutarate Pathway	*B. subtilis*	Pharmaceutical	Disabling the KG-A utilization pathway	12.21 g/L	[42]
21	γ-aminobutyric Acid	L-amino Acid Deaminase Pathway	*B. subtilis*	Pharmaceutical	Glutamate decarboxylase (GAD) Overexpression	5.26 g/L	[53]
22	Polyhydroxy Butyrate	β-oxidation Pathway	*Pseudomonas putida* and *P. entomophila*	Bioplastic	*fadA* & *fadB* were tried to block the β-oxidation pathway		[60]
23	Polyhydroxy Alkanoates	β-oxidation Pathway	*E. coli*	Bioplastic	Deletion native succinate semialdehyde dehydrogenase genes (*sad* & *gabD*)		[60]
24	4-Hydroxytetradecanoate	β-oxidation Pathway	*P. entomophila* LAC23	Bioplastic	Deleted β-oxidation pathway		[62]

References

1. Chauhan, S., Velramar, B., Soni, R.K., Mishra, M., and Sudheer, P.D.V.N., *Biofuels: Sources, Modern Technology Developments and Views on Bioenergy Management*, in *Biotechnology for Biofuels: A Sustainable Green Energy Solution*, N. Kumar, Editor. 2020, Springer Singapore: Singapore. p. 197-219.
2. Dien, B.S., Cotta, M.A., and Jeffries, T.W., Bacteria engineered for fuel ethanol production: current status. *Appl Microbiol Biotechnol.* 63, 258-66, 2003.
3. Sudheer, P.D.V.N., Chauhan, S., and Velramar, B., *Bio-Hydrogen: Technology Developments in Microbial Fuel Cells and Their Future Prospects*, in *Biotechnology for Biofuels: A Sustainable Green Energy Solution*, N. Kumar, Editor. 2020, Springer Singapore: Singapore. p. 61-94.
4. Krishnan, A., Mcneil, B.A., and Stuart, D.T., Biosynthesis of Fatty Alcohols in Engineered Microbial Cell Factories: Advances and Limitations. *Frontiers in Bioengineering and Biotechnology*. 8, 2020.
5. Mishra, M., Chauhan, S., Velramar, B., Soni, R.K., and Pamidimarri, S.D.V.N., Facile bioconversion of vegetable food waste into valuable organic acids and green fuels using synthetic microbial consortium. *Korean Journal of Chemical Engineering*. 38, 833-842, 2021.
6. Derkx, P.M., Janzen, T., Sørensen, K.I., Christensen, J.E., Stuer-Lauridsen, B., and Johansen, E., The art of strain improvement of industrial lactic acid bacteria without the use of recombinant DNA technology. *Microb Cell Fact.* 13 Suppl 1, S5, 2014.
7. Majidian, P., Tabatabaei, M., Zeinolabedini, M., Naghshbandi, M.P., and Chisti, Y., Metabolic engineering of microorganisms for biofuel production. *Renewable and Sustainable Energy Reviews*. 82, 3863-3885, 2018.
8. Peralta-Yahya, P., Zhang, F., Cardayre, S., and Keasling, J., Microbial engineering for the production of advanced Biofuels. *Nature*. 488, 320-8, 2012.
9. Ahn, S.Y., Jang, S., Sudheer, P., and Choi, K.Y., Microbial Production of Melanin Pigments from Caffeic Acid and L-tyrosine Using Streptomyces glaucescens and FCS-ECH-Expressing Escherichia coli. *Int J Mol Sci.* 22, 2021.
10. Sudheer, P.D.V.N., Yun, J., Chauhan, S., Kang, T.J., and Choi, K.-Y., Screening, expression, and characterization of Baeyer-Villiger monooxygenases for the production of 9-(nonanoyloxy)nonanoic acid from oleic acid. *Biotechnology and Bioprocess Engineering*. 22, 717-724, 2017.
11. Sudheer, P.D.V.N., Seo, D., Kim, E.-J., Chauhan, S., Chunawala, J.R., and Choi, K.-Y., Production of (Z)-11-(heptanoyloxy)undec-9-enoic acid from ricinoleic acid by utilizing crude glycerol as sole carbon source in engineered Escherichia coli expressing BVMO-ADH-FadL. *Enzyme and Microbial Technology*. 119, 45-51, 2018.
12. Becker, J., Lange, A., Fabarius, J., and Wittmann, C., Top value platform chemicals: bio-based production of organic acids. *Current Opinion in Biotechnology*. 36, 168-175, 2015.

13. Robak, K. and Balcerek, M., Current state-of-the-art in ethanol production from lignocellulosic feedstocks. *Microbiological Research*. 240, 126534, 2020.
14. Ingram, L.O., Conway, T., Clark, D.P., Sewell, G.W., and Preston, J.F., Genetic engineering of ethanol production in Escherichia coli. *Applied and environmental microbiology*. 53, 2420-2425, 1987.
15. Boock, J.T., Freedman, A.J.E., Tompsett, G.A., Muse, S.K., Allen, A.J., Jackson, L.A., Castro-Dominguez, B., Timko, M.T., Prather, K.L.J., and Thompson, J.R., Engineered microbial biofuel production and recovery under supercritical carbon dioxide. *Nature Communications*. 10, 587, 2019.
16. Mukhopadhyay, A., Tolerance engineering in bacteria for the production of advanced biofuels and chemicals. *Trends in Microbiology*. 23, 498-508, 2015.
17. Fithriani, Suryadarma, P., and Mangunwidjaja, D., Metabolic Engineering of Escherichia coli Cells for Ethanol Production under Aerobic Conditions. *Procedia Chemistry*. 16, 600-607, 2015.
18. Mohagheghi, A., Evans, K., Chou, Y.C., and Zhang, M., Cofermentation of glucose, xylose, and arabinose by genomic DNA-integrated xylose/arabinose fermenting strain of Zymomonas mobilis AX101. *Appl Biochem Biotechnol*. 98-100, 885-98, 2002.
19. Linger, J.G., Adney, W.S., and Darzins, A., Heterologous Expression and Extracellular Secretion of Cellulolytic Enzymes by Zymomonas mobilis. *Applied and Environmental Microbiology*. 76, 6360, 2010.
20. Kim, Y., Ingram, L.O., and Shanmugam, K.T., Construction of an Escherichia coli K-12 mutant for homoethanologenic fermentation of glucose or xylose without foreign genes. *Appl Environ Microbiol*. 73, 1766-71, 2007.
21. Kuyper, M., Hartog, M.M.P., Toirkens, M.J., Almering, M.J.H., Winkler, A.A., Van Dijken, J.P., and Pronk, J.T., Metabolic engineering of a xylose-isomerase-expressing Saccharomyces cerevisiae strain for rapid anaerobic xylose fermentation. *FEMS Yeast Research*. 5, 399-409, 2005.
22. Matsushika, A., Inoue, H., Kodaki, T., and Sawayama, S., Ethanol production from xylose in engineered Saccharomyces cerevisiae strains: current state and perspectives. *Applied Microbiology and Biotechnology*. 84, 37-53, 2009.
23. Vallari, R.C., Cook, W.J., Audino, D.C., Morgan, M.J., Jensen, D.E., Laudano, A.P., and Denis, C.L., Glucose repression of the yeast ADH2 gene occurs through multiple mechanisms, including control of the protein synthesis of its transcriptional activator, ADR1. *Molecular and cellular biology*. 12, 1663-1673, 1992.
24. Ye, W., Zhang, W., Liu, T., Tan, G., Li, H., and Huang, Z., Improvement of Ethanol Production in Saccharomyces cerevisiae by High-Efficient Disruption of the ADH2 Gene Using a Novel Recombinant TALEN Vector. *Frontiers in Microbiology*. 7, 2016.
25. Atsumi, S., Cann, A.F., Connor, M.R., Shen, C.R., Smith, K.M., Brynildsen, M.P., Chou, K.J., Hanai, T., and Liao, J.C., Metabolic engineering of Escherichia coli for 1-butanol production. *Metab Eng*. 10, 305-11, 2008.

26. Bennett, G. and Rudolph, F., The central metabolic pathway from acetyl-CoA to butyryl-CoA in. *FEMS Microbiology Reviews*. 17, 241-249, 2006.
27. Jiang, Y., Xu, C., Dong, F., Yang, Y., Jiang, W., and Yang, S., Disruption of the acetoacetate decarboxylase gene in solvent-producing Clostridium acetobutylicum increases the butanol ratio. *Metabolic Engineering*. 11, 284-291, 2009.
28. Han, B., Gopalan, V., and Ezeji, T.C., Acetone production in solventogenic Clostridium species: new insights from non-enzymatic decarboxylation of acetoacetate. *Applied Microbiology and Biotechnology*. 91, 565-576, 2011.
29. Tummala, S.B., Welker, N.E., and Papoutsakis, E.T., Design of Antisense RNA Constructs for Downregulation of the Acetone Formation Pathway of Clostridium acetobutylicum. *Journal of Bacteriology*. 185, 1923, 2003.
30. Jang, Y.-S., Lee, J.Y., Lee, J., Park, J.H., Im, J.A., Eom, M.-H., Lee, J., Lee, S.-H., Song, H., Cho, J.-H., Seung, D.Y., and Lee, S.Y., Enhanced Butanol Production Obtained by Reinforcing the Direct Butanol-Forming Route in Clostridium acetobutylicum. *mBio*. 3, e00314-12, 2012.
31. Hou, X., Peng, W., Xiong, L., Huang, C., Chen, X., Chen, X., and Zhang, W., Engineering Clostridium acetobutylicum for alcohol production. *Journal of Biotechnology*. 166, 25-33, 2013.
32. George, H.A., Johnson, J.L., Moore, W.E., Holdeman, L.V., and Chen, J.S., Acetone, Isopropanol, and Butanol Production by Clostridium beijerinckii (syn. Clostridium butylicum) and Clostridium aurantibutyricum. *Appl Environ Microbiol*. 45, 1160-3, 1983.
33. Walther, T. and François, J.M., Microbial production of propanol. *Biotechnology Advances*. 34, 984-996, 2016.
34. Jojima, T., Inui, M., and Yukawa, H., Production of isopropanol by metabolically engineered Escherichia coli. *Applied Microbiology and Biotechnology*. 77, 1219-1224, 2008.
35. Hanai, T., Atsumi, S., and Liao, J.C., Engineered Synthetic Pathway for Isopropanol Production in Escherichia coli. *Applied and Environmental Microbiology*. 73, 7814, 2007.
36. Inokuma, K., Liao, J.C., Okamoto, M., and Hanai, T., Improvement of isopropanol production by metabolically engineered Escherichia coli using gas stripping. *Journal of Bioscience and Bioengineering*. 110, 696-701, 2010.
37. Soma, Y., Tsuruno, K., Wada, M., Yokota, A., and Hanai, T., Metabolic flux redirection from a central metabolic pathway toward a synthetic pathway using a metabolic toggle switch. *Metabolic Engineering*. 23, 175-184, 2014.
38. Tamakawa, H., Mita, T., Yokoyama, A., Ikushima, S., and Yoshida, S., Metabolic engineering of Candida utilis for isopropanol production. *Applied Microbiology and Biotechnology*. 97, 6231-6239, 2013.
39. Dellomonaco, C., Clomburg, J.M., Miller, E.N., and Gonzalez, R., Engineered reversal of the β-oxidation cycle for the synthesis of fuels and chemicals. *Nature*. 476, 355-359, 2011.
40. Lennen, R.M. and Pfleger, B.F., Microbial production of fatty acid-derived fuels and chemicals. *Current opinion in biotechnology*. 24, 1044-1053, 2013.

41. Steen, E.J., Kang, Y., Bokinsky, G., Hu, Z., Schirmer, A., Mcclure, A., Del Cardayre, S.B., and Keasling, J.D., Microbial production of fatty-acid-derived fuels and chemicals from plant biomass. *Nature*. 463, 559-562, 2010.
42. Zhang, K., Sawaya, M.R., Eisenberg, D.S., and Liao, J.C., Expanding metabolism for biosynthesis of nonnatural alcohols. *Proceedings of the National Academy of Sciences*. 105, 20653, 2008.
43. Li, Y., Yang, S., Ma, D., Song, W., Gao, C., Liu, L., and Chen, X., Microbial engineering for the production of C2–C6 organic acids. *Natural Product Reports*. 2021.
44. Causey, T.B., Zhou, S., Shanmugam, K.T., and Ingram, L.O., Engineering the metabolism of Escherichia coli W3110 for the conversion of sugar to redox-neutral and oxidized products: homoacetate production. *Proc Natl Acad Sci U S A*. 100, 825-32, 2003.
45. Chang, D.-E., Jung, H.-C., Rhee, J.-S., and Pan, J.-G., Homofermentative Production of Lactate in Metabolically Engineered Escherichia coli RR1. *Applied and Environmental Microbiology*. 65, 1384, 1999.
46. Dien, B.S., Nichols, N.N., and Bothast, R.J., Recombinant Escherichia coli engineered for production of L-lactic acid from hexose and pentose sugars. *J Ind Microbiol Biotechnol*. 27, 259-64, 2001.
47. Mazumdar, S., Blankschien, M.D., Clomburg, J.M., and Gonzalez, R., Efficient synthesis of L-lactic acid from glycerol by metabolically engineered Escherichia coli. *Microb Cell Fact*. 12, 7, 2013.
48. Kim, P., Laivenieks, M., Vieille, C., and Zeikus, J.G., Effect of Overexpression of Actinobacillus succinogenes Phosphoenolpyruvate Carboxykinase on Succinate Production in Escherichia coli. *Applied and Environmental Microbiology*. 70, 1238, 2004.
49. Sánchez, A.M., Bennett, G.N., and San, K.Y., Novel pathway engineering design of the anaerobic central metabolic pathway in Escherichia coli to increase succinate yield and productivity. *Metab Eng*. 7, 229-39, 2005.
50. Sospedra, I., De Simone, C., Soriano, J.M., Mañes, J., Ferranti, P., and Ritieni, A., Liquid chromatography-ultraviolet detection and quantification of heat-labile toxin produced by enterotoxigenic E. coli cultured under different conditions. *Toxicon*. 141, 73-78, 2018.
51. Park, S.A., Bhatia, S.K., Park, H.A., Kim, S.Y., Sudheer, P., Yang, Y.H., and Choi, K.Y., Bacillus subtilis as a robust host for biochemical production utilizing biomass. *Crit Rev Biotechnol*. 1-34, 2021.
52. Hossain, G.S., Li, J., Shin, H.D., Liu, L., Wang, M., Du, G., and Chen, J., Improved production of α-ketoglutaric acid (α-KG) by a Bacillus subtilis whole-cell biocatalyst via engineering of L-amino acid deaminase and deletion of the α-KG utilization pathway. *J Biotechnol*. 187, 71-7, 2014.
53. Zhang, C., Lu, J., Chen, L., Lu, F., and Lu, Z., Biosynthesis of γ-aminobutyric acid by a recombinant Bacillus subtilis strain expressing the glutamate decarboxylase gene derived from Streptococcus salivarius ssp. thermophilus Y2. *Process Biochemistry*. 49, 1851-1857, 2014.

54. Chevreul, M.E., List, G.R., and Wisniak, J., A chemical study of oils and fats of animal origin. 2009, St. Eutrope-de Born, France; Urbana, IL: Sàrl Dijkstra-Tucker Carbougnères ; Distributed by AOCS Press.
55. Choi, K.H., Heath, R.J., and Rock, C.O., beta-ketoacyl-acyl carrier protein synthase III (FabH) is a determining factor in branched-chain fatty acid biosynthesis. *J Bacteriol.* 182, 365-70, 2000.
56. Rowland, O. and Domergue, F., Plant fatty acyl reductases: Enzymes generating fatty alcohols for protective layers with potential for industrial applications. *Plant Science.* 193-194, 28-38, 2012.
57. Akhtar, M.K., Turner, N.J., and Jones, P.R., Carboxylic acid reductase is a versatile enzyme for the conversion of fatty acids into fuels and chemical commodities. *Proceedings of the National Academy of Sciences.* 110, 87, 2013.
58. Ashter, S.A., *2 - Overview of Biodegradable Polymers*, in *Introduction to Bioplastics Engineering*, S.A. Ashter, Editor. 2016, William Andrew Publishing: Oxford. p. 19-30.
59. Li, Z., Yang, J., and Loh, X.J., Polyhydroxyalkanoates: opening doors for a sustainable future. *NPG Asia Materials.* 8, e265-e265, 2016.
60. Sánchez, R.J., Schripsema, J., Da Silva, L.F., Taciro, M.K., Pradella, J.G.C., and Gomez, J.G.C., Medium-chain-length polyhydroxyalkanoic acids (PHAmcl) produced by Pseudomonas putida IPT 046 from renewable sources. *European Polymer Journal.* 39, 1385-1394, 2003.
61. Chen, G.-Q., Hajnal, I., Wu, H., Lv, L., and Ye, J., Engineering Biosynthesis Mechanisms for Diversifying Polyhydroxyalkanoates. *Trends in Biotechnology.* 33, 565-574, 2015.
62. Li, Z.-J., Shi, Z.-Y., Jian, J., Guo, Y.-Y., Wu, Q., and Chen, G.-Q., Production of poly(3-hydroxybutyrate-co-4-hydroxybutyrate) from unrelated carbon sources by metabolically engineered Escherichia coli. *Metabolic Engineering.* 12, 352-359, 2010.

4

Production of Biomethane and Its Perspective Conversion: An Overview

Rajesh K. Srivastava[1]* and Prakash Kumar Sarangi[2]

[1]*Department of Biotechnology, GIT, GITAM (Deemed to be University), Visakhapatnam, A. P. India*
[2]*College of Agriculture, Central Agricultural University, Imphal, India*

Abstract
There is great concern about the generation of renewable sources so as to mitigate the excessive use of conventional fossil fuels. The latter not only create environmental impacts, but will also diminish in the coming future. Biomethane can be generated by anaerobic digestion methods from various waste biomass. Microbial processes are employed for generation of biomethane from a wide range of waste biomass. Biomethane can also be converted into biomethanol by microbial action. Biomethanol can also be produced from waste biomass by other modes, such as thermochemical. This chapter gives an overview of production of biomethane and subsequent biomethanol.

Keywords: Biomethanol, biomethane, waste biomass, fossil fuels

4.1 Introduction

There are many sources, such as production and transport coal, natural gas, and deep petroleum oil sources that are found to emit methane gas in the environment. Further, methane can be emitted from livestock and agricultural practices, land use, and by decay of organic wastes in municipal solid waste landfills. Release of methane into the environment is reported from cows and bogs. It was found that most human activity that can drive up the levels of destructive greenhouse gases [1]. Total methane emission (nearly

**Corresponding author*: rajeshksrivastava73@yahoo.co.in

Prakash Kumar Sarangi (ed.) Biorefinery Production of Fuels and Platform Chemicals, (93–104) © 2023 Scrivener Publishing LLC

172 to 195 teragram methane per year) in partition location of total fossil fuels was found due to human induced activity and natural processes from geological sources such as seeps and mud volcanoes. In recent debates, this methane emission accounted for about 40 to 60 teragrams methane/year. Geological activities have contributed less than 15.4 teragrams of methane each year from the end of Pleistocene periods [2].

Plants can die in bogs or fens (as reported as two of the wetlands) via sinking into acidic environments and low oxygen environments. In this location, microbial activity cannot break them down as well as in other habitats. So, partial decomposition of plants can accumulate and create peat (brown deposit), which looks like soil structure. Its carbon content allows it to burn and methane can be emitted. Further, peat wetland, known as peatlands, [2] show the capacity to store twice as much, compared to an area with a greater quantity of carbon, such as a forest (despite of covering 3% of lands). The most common sources are found across the North America and Europe are involved in removing muddy peat that can be used for fertilizer agents [3].

Next, some countries are found to burn peat for heating in homes and for business tasks. When a cow burps or passes gas, it can contribute a little puff of methane into the atmosphere. It has reported that bongs and fens are more acidic than other types of wetland and naturally provide 7% global methane emissions. These two sources have shown a difference between the two peat habitats that can receive water. Stream flow free through fens, while bogs are dependent on rainfall to replenish water [2, 3].

Methane fluxes have evaluated under water as stable in mesocosms planted with *Asclepias incarnata* L. and mesocosms planted with *Alisma triviale* pursh. A mesocosm can bring a part of the natural environment into controlled conditions. Researchers have tested interactive effects in saturated and unsaturated restored field locations and methane fluxes from the plant and the surrounding soil. Methane fluxes studied in mesocosms planted with *Asclepias incarnata* found 8 times more than the controlled condition (i.e., in plant) or *Alisma trivial* [4, 5].

Further, *Alisma trivial* mesocosms showed a lower carbon dioxide to methane ratio (due to less methanogenic species dominance) than control mesocosms, but this plant didn't show much difference compared to *Asclepias incarnate* mesocosms [6, 7]. In this context, hydrology can be the dominant structure for controlling methane flux in both plant species that shown 10 times more methane production in saturated plots than unsaturated plots. Next, hydrology can be added with plant species composition and can better predict methane fluxes from wetlands soils and it can aid in designing restored wetlands that can minimize GHGs emissions [4, 6].

A puff coming out from cow plumbing can have a big effect on climatic change due to methane emission (28 times more powerful than carbon dioxide in warming of earth). In the last 20 years, more than 80 times power methane emission in our atmosphere are reported from the industrial revolution and it has contributed 20% more warming to the planet [8]. This can impact the atmosphere 200 times less than carbon dioxide [9]. This gas is abundant and one of the more dangerous gases of the GHGs. Further chemical structures are reported to trap heat that is experienced by a warmed planet. Methane is a simple gas, made up of single carbon atoms with four atoms of hydrogen [10]. It was shown as fleeting compared to other GHG gases like CO_2. Once it is spewed into the atmosphere, it showed lasting capability of about a decade before its cycled out. There are many sources of CH_4 that can load into the atmosphere with a constantly regenerating or increasing nature [1, 11].

4.1.1 Sources of Methane

In the current period, 60% methane has been contributed to the atmosphere that comes from multiples sources and human activities that started to influence the carbon cycle in dramatic or adverse ways. As discussed earlier in this paper, that natural emission of methane is generated or comes from soggy sources such as wetlands (different types of bogs) [7]. The microbes system in mammal guts catabolize or degrade the eaten organic material by splitting it into CO_2. But, such microbes can still live in oxygen deprived spots like wetland soil (waterlogged conditions) and are shown to produce the methane leaked into atmosphere. These types of locations on earth can contribute nearly 1/3rd of the methane that can float in the modern atmosphere via wetland activity [12, 13].

Other natural methane emission sources are grounds near some oil and gas deposit locations and also the mouths of volcanoes. Further methane can leak out of thawing permafrost in the Arctic and builds up in the sediments near sea locations. This can waft away from burning landscapes and contribute atmospheric CO_2. Next, methane can be produced by termites via chewing piles of woody detritus. These other natural sources, excluding wetlands, can contribute 10% of the total methane emissions each year [14]. Researchers have worked on the variability of composition that was observed between sites with an ethane to methane ratio of 5:3. In this regard, 13C and D methane isotopic composition was found around 40 and 240% (per mille sign), respectively [6].

Offshore plume was found to more spatially narrow than the expectations of the plume width based on terrestrial stability. A modified Gaussian dispersion methodology (GDM) was used with empirical manner measured horizontal plume width as an estimation of the emission rate of methane. Nearly 103 sites were studied in this regard that included shallow and deepwater offshore platforms and drill-ships. It found range of 0 to 190kg/h with 95% confidence limits estimation at a factor of 10 [15]. 20% of the total methane emissions in all the sample sites are found using two emitters that skewered with observed distribution. It has greater throughput of the deepwater facilities with moderate emission rates compared to shallow water sites [16].

4.1.2 Methane from Human Activity

There are many reports that have discussed the human induced activities for bulk quantity of methane in our atmosphere. Cows and other grazing mammals are shown to have methane producing belches and it can show releasing capacities. These sources can provide a host for microbes including its stomach and can work as gut filling hitchhikers that help in breakdown of complex organic matters and absorbing nutrients from tough grasses [16]. These microbial activities are shown to produce methane as waste and wafts out of the both ends of cows. Further, manure from cattle and other grazer animals can also provide a site for microbes performing the business or tasks of methane production that come from nearly 1.4 billion cattle in the world. Next, the increasing demand of beef and dairy has caused the number of grazing animals to contribute 40% of the annual methane budget [15].

Next, agricultural processes can contribute methane into the atmosphere. Rice paddies can contribute more percentage than wetlands. Flooded rice paddies with calm water are found to be low in O_2 content and can be natural home for methane producing bacteria. Rice production took off in Asia nearly 500 years ago. Later, methane was recorded as a tiny bubble of ancient air trapped in an ice core in Antarctica and it rose rapidly via methane impact on climate [15, 16].

4.1.3 Impact of Methane on Climatic Change and Future

As we have noticed, methane has impacted rapid warming events deep in the Earth and it is shown in ancient periods. It occurred under high pressures that are found in deep at the bottom of the ocean and there, methane are reported in solid form (i.e., slush-like material), also known as methane hydrate. This is frozen in vast quantities at the bottom of the sea in the chemical state [6]. This methanol hydrate is found in a stable form, but can be disturbed by a plume of warm water. Massive warming events occurred 55 billion years ago and it has shown to be kicked off by destabilized methanol hydrates. In this process, methane percolation up from seafloor is found in the atmosphere. Further, methane as a heat trapping gas is forcing the planet to warm in a drastic and quick manner. The modern atmosphere is reported to contain a methane concentration more than 150% and from 1750, this level still continues to rise [14].

4.1.4 Advancements and Challenges

The release of methane into the environment comes from cows and bogs, as well as from most human activity. It can drive up the levels of destructive greenhouse gases [1]. Total methane emission (nearly 172 to 195 teragram methane per year) in partially from fossil fuels and is also due to human induced activity and natural processes (geological sources such as seeps and mud volcanoes). Geological activities (less than 15.4 teragram methane each year) are reported from the end of Pleistocene periods [1]. Other natural methane emission sources are grounds near some oil and gas deposit locations and also mouths of volcanoes. Further methane can leak out of thawing permafrost in the Arctic and builds up in the sediments under shallow sea locations. This can waft away from burning landscapes and contribute atmospheric CO_2. Variability of composition that was observed between sites has an ethane to methane ratio of 5:3. In this regard, 13C and D methane isotopic composition was found around 40 and 240‰ (per mille sign), respectively [6]. This is a big challenge for the huge quantity of methane in our environment and it needs to transform into other useful products like methanol or other products.

Puffs coming out from cow plumbing are reported as a big effect on climatic change due to methane emissions (28 times more powerful than carbon dioxide in warming of earth). In the last 20 years, more than 80 times of the methane emissions in our atmosphere are reported from the industrial revolution and it has contributed 20% more warming on the earth [8]. This can impact the concentrated atmosphere 200 times less than carbon dioxide [9]. Also, the microbe systems in mammal guts catabolize or degrade organic material that is splitting into CO_2. But, such microbes can still live in oxygen deprived spots like wetland soil (waterlogged conditions) are shown to produce methane that leaks into the atmosphere. These types of locations in the earth can contribute nearly

1/3rd of methane that can float in the modern atmosphere via wetland activity [12, 13]. Other natural methane emission sources are grounds near some oil and gas deposit locations and also mouths of volcanoes. Further methane can leak out of thawing permafrost in the Arctic and builds up in the sediments near sea locations. This can waft away from burning landscapes, contributing atmospheric CO_2. Next, methane can b produces by termites via chewing piles of woody detritus [14]. Human induced activities for bulk quantity of methane are also reported in our atmosphere. Also, cows and others grazing mammals have methane producing belches that can show releasing capacities. These sources can provide a host for microbes including its stomach and can work as gut filling hitchhikers via helping in breakdown of complex organic matters while absorbing nutrient from tough grasses [16].

As we know, methane molecules are found in perfect tetrahedron form with four C-H bonds in uniform nature. Limitation of this process is the oxidation of CO and H_2 that can result in wanted CO_2 and water. Huge quantities of heat are also wasted during this process. In modern plants, this efficiency of conversion can increase by using released heat for steam reforming tasks and partial oxidized methane is reported with O_2 and reform steam [17, 18]. General syngas production processes have shown different operating designs and conditions with economic limitations. In methanol production process systems, the rate of H_2/CO was found to be more stable (2-3 times more). This process regulation was found to be very problematic nature [19]. Now, there are a number of conversion technologies that are used for conversion of methane into methanol.

Further, methane can show operating problems or issues (cocking) and it can reduce the economic efficiency of methanol production from synthesis gas as a feed [19, 20]. Methanol production occurs in a tow reactor, but it can increase process problems. Also, it is due to an increases number of process steps that need more space and equipment for methanol production. Two systems are discussed for controlling the process via integrating processes in desirable levels and simple form and from ancient periods onwards, active research programs are found around the world that promote direct conversion of methane into methanol [21]. Methanol synthesis by direct hydrogenation of CO_2 molecules has had a number of analyses is done. Methanol production is reported from a current standard approach using synthesis gas. This approach is found to require a huge amount of energy with high rates of emission of carbon dioxides. The direct method is not found for carbon dioxide emission, but it is a good process with a desirable yield of methanol [22]. Optimal operating conditions for methanol production from hydrogenation of carbon dioxide are discussed and it showed minimum production cost. This approach has used response surface methodology (RSM) for an optimal or high yield of methanol production from a carbon dioxide hydrogenation approach [23].

This approach can show a disadvantage for required electricity supply. We can get inexpensive electricity from natural energy as hydroelectric power, wind power, and geothermal power. This approach of electricity generation can help in reduction of methanol production. Nowadays, people use hydro-power as an ideal approach for electricity, but it needs the stability of the seasons of the year. This approach to electricity generation is used for methanol production (4000t/year) via hydrogenation of carbon dioxide. Methanotrophic bacteria can oxidize the methane to CO_2 via utilization of sequential reactions that were catalyzed by series of enzymes, such as methanooxigenase (MMO), methanol dehydrogenase (MDH), formaldehyde dehydrogenase (FDH), and formate dehydrogenase. In this regard, a suspension of methanotrophic bacteria (*methylosinus trichosporium* IMV 1011) were used and incubated at 32^{0}C with methane and oxygen [24]. Methanol production by carbon dioxide has shown many advantages such as enhanced environmental sustainability of process, decreased CO_2 emissions, mild process operating conditions, and suitable methanol production rates with high selectivity [25, 26]. But, plasma-based reforming is one of the advanced and effective approaches for methanol production. This approach to production of methanol can be completed in a single step and is better than the traditional approach. Now, researchers are reported to apply the traditional plasma-based dry reforming and also plasma based multi-reforming [27].

Finally, from all these approaches, methanol generation and its utilization can offer advantages as a liquid fuel and also as a building block for industrial chemical synthesis tasks. A plasma assisted approach or technology can provide a path of transition to a mixed carbon-hydrogen economy with a reduction in net carbon dioxide emissions and also pushes for a carbon dioxide neutral world. Carbon dioxide emissions from road, water, and transport tasks can be found to diffuse and high concentrations in our healthy atmosphere can be only solved via development of technologies for capturing CO_2 from the air. This gas can be utilized for final conversion into methanol [28, 29]. The photo-catalyst technique is reported to enhance productivities and yield (3.3 times) via increasing the reactant temperature from 93^{0}C to 260^{0}C via use of a constant surface temperature (350 °C). This photo-catalyst is reported to utilize the Cu^+ and this catalyst played a critical role in thermal catalysis by utilizing its active sites for product formation (i.e., methanol) [30]. In current developed designs, the most feasible design for methanol production is needed for an energy requirement (21.9GJ) and product costs ($ 142.5/ton). A proposed coal to methanol process, assisted externally by hydrogen from an electrolysis plant can achieve higher production rates and low carbon dioxide emission. These processes cannot process commercially due to higher costs of hydrogen [31, 32]. Solar assisted methanol production is reported from direct utilization of carbon dioxide and hydrogen gas and is also found as a very important approach for the development of the energy economy. Numbers of growing activity of humans are reported and photo-catalysts is reported as an efficient and stable process for hydrogenating gaseous carbon dioxide into methanol at ambient pressure in a highly selective process [33, 34]. There is the big challenge of hydrogen cost and this can depend on resource type and production approach. Now, people are using hydrogen from renewable resources that can be used in reforming of methanol processes and it was continuously operated in a tubular reactor made up of a metal nickel-based alloy Inconel 625. For this approach, experiments were carried out at pressure from 25 to 45MPa and temperature in range of 400 to 600^{0}C [35].

References

1. Hmiel, B., Petrenko, V.V., Dyonisius, M.N., Buizert, C., Smith, A.M., Place, P.F., Harth, C., Beaudette, R., Hua, Q., Yang, B., Vimont, I., Michel, S.E, Severinghaus, J.P., Etheridge, D., Bromley, T., Schmitt, J., Faïn, X., Weiss, R.F., Dlugokencky, E., Preindustrial 14CH4 indicates greater anthropogenic fossil CH4 emissions. Nat., 2020, 578 (7795), 409 DOI: 10.1038/s41586-020-1991-8
2. Silvey, C., Jarecke, K.M., Hopfensperger, K., Loecke, T.D., Burgin. A.J., Plant Species and Hydrology as Controls on Constructed Wetland Methane Fluxes. Soil Sci. Soc.Amer. J., 2019, 83(3), 848-855. DOI: 10.2136/sssaj2018.11.0421.
3. Laanbroek, H.J. Methane emission from natural wetlands: interplay between emergent macrophytes and soil microbial processes. A mini-review. Ann Bot., 2010, 105(1), 141–153. doi: 10.1093/aob/mcp201
4. Carmichael, M.J., Bernhardt, E.S., Bräuer, S.L., Smith, W.K. The role of vegetation in methane flux to the atmosphere: should vegetation be included as a distinct category in the global methane budget? Biogeochem., 2014, 119, 1–24. https://doi. org/ 10.1007/s10533-014-9974-1
5. Smith, G.J., Angle, J.C., Solden, L.M., Borton, M.A., Morin, T.H., Daly, R.A., Johnston, M.D., Stefanik, K.C., Wolfe, R., Gil, B., Wrighton, K.C., Members of the Genus Methylobacter Are Inferred To Account for the Majority of

Aerobic Methane Oxidation in Oxic Soils from a Freshwater Wetland. mBio., 2018, 9(6), e00815-18. doi: 10.1128/mBio.00815-18.

6. Riddick S.N., Mauzerall, D.L., Celia M., Harris, N.R.P., Allen, G., Pitt, J., Staunton-Sykes J., Forster, G.L., Kang, M., Lowry, D., Nisbet E.G., Manning, A.J., Methane emissions from oil and gas platforms in the North Sea. Atmos. Chem. Phys., 2019, 19, 9787–9796, https://doi.org/10.5194/acp-19-9787-2019.
7. Covey, K.R., Megonigal, J.P., Methane production and emissions in trees and forests. New Phytologist., 2019, 222, 35–51. https://doi.org/10.1111/nph.15624C
8. Nara, H., Tanimoto, H., Tohjima, Y., Mukai, H., Nojiri, Y., and Machida, T.: Emissions of methane from offshore oil and gas platforms in Southeast Asia, Sci. Rep., 2015, 4, 6503, https://doi.org/10.1038/srep06503
9. Johnson, M. R., Tyner, D.R., Conley, S., Schwietzke, S., Zavala-Araiza, D. Comparisons of Airborne Measurements and Inventory Estimates of Methane Emissions in the Alberta Upstream Oil and Gas Sector, Environ. Sci. Technol., 2017, 51, 13008–13017. https://doi. org /10.1021/acs.est.7b03525.
10. Mahmudah, W., Noviasari, S.A., Damayanti, D., Prihantono, G., Estimated Methane Gas Emissions (CH4) from the Utilization of a Biogas Technology as an Alternative Fuel Source in Ploso Ngamban Village, Kendal, Ngawi East Java. Sys Rev. Pharm., 2020, 11(6), 1373-1375. doi: 10.31838/srp.2020.6.198.
11. Ma, K., Ma, A., Zheng, G, Ren, G., Xie, F., Zhou, H., Yin, J., Liang, Y., Zhuang, X., Zhuang, G., Mineralosphere Microbiome Leading to Changed Geochemical Properties of Sedimentary Rocks from Aiqigou Mud Volcano, Northwest China. Microorgan., 2021, 9(3), 560. doi: 10.3390/microorganisms9030560
12. Yip, D.Z., Veach, A.M., Yang, Z.K., Cregger, M.A., Schadt, CW., Methanogenic Archaea dominate mature hardwood habitats of Eastern Cottonwood (Populusdeltoides). New Phytolog., 2018, 222, 115–122. https://doi.org/10.1111/nph.15346
13. You, Y., Staebler, R.M., Moussa, S.G., Beck, J., Mittermeier R. L.,, Methane emissions from an oil sands tailings pond: a quantitative comparison of fluxes derived by different methods, Atmos. Meas. Tech., 2021, 14, 1879–1892. https://doi.org/10.5194/amt-14-1879-2021.
14. Yacovitch, T.I., Daube, C., Herndon S.C., Methane Emissions from Offshore Oil and Gas Platforms in the Gulf of Mexico. Environ. Sci. Technol., 2020, 54(6), 3530–3538. https://doi.org/10.1021/acs.est.9b07148.
15. Tuccella, P., Thomas, J. L, Law, K. S., Raut, J.-C., Marelle, L., Roiger, A., Weinzierl, B., Denier van der Gon, H.A.C., Schlager, H., Onishi, T., Air pollution impacts due to petroleum extraction in the Norwegian Sea during the ACCESS aircraft campaign. Elem. Sci. Anth., 2017, 5, 25. DOI: 10.1525/elementa.124.
16. Lee, J.D., Mobbs, S. D., Wellpott, A., Allen, G., Bauguitte, S.J.-B., Burton, R.R., Camilli, R., Coe, H., Fisher, R.E., France, J.L., Gallagher, M., Hopkins, J. R., Lanoiselle, M., Lewis, A.C., Lowry, D., Nisbet, E.G., Purvis, R.M., O'Shea,

S., Pyle, J.A., Ryerson, T.B., Flow rate and source reservoir identification from airborne chemical sampling of the uncontrolled Elgin platform gas release. Atmos. Meas. Tech., 2018, 11, 1725- 1739, DOI: 10.5194/amt-11-1725-2018

17. Zhang C, Jun KW, Gao R, Kwak, G., Park, H.-G., Carbon dioxide utilization in a gas-to-methanol process combined with CO2/Steam-mixed reforming: Techno-economic analysis. Fuel., 2017, 190, 303–311. https://doi.org/10.1016/j. fuel.2016. 11.008
18. Yang, Y., Liu, J., Shen, W., Li, J., Chien, I-L., High-efficiency utilization of CO_2 in the methanolproduction by a novel parallel-series system combining steam and dry methane reforming. Ener., 2018, 158, 820–829 https://doi.org/10.1016/j .energy. 2018.06.061
19. van de Water, L.G.A., Wilkinson, S.K., Smith, R.A.P., Watson, M.J., Understanding methanol synthesis from CO/H2 feeds over Cu/CeO2 catalysts. J. Catalys., 2018, 364, 57-68. https://doi.org/10.1016/j.jcat.2018.04.026
20. Kajaste, R., Hurme, M., Oinas, P., Methanol-Managing greenhouse gas emissions in the production chain by optimizing the resource base. AIMS Ener., 2018, 6(6), 1074–1102. DOI: 10.3934/energy.2018.6.1074
21. Shindell, D., Borgford-Parnell, N., Brauer, M., Haines, A., Kuylenstierna, J.C.I., Leonard, S.A., Ramanathan, V., Ra, A., A climate policy pathway for near- and long-term benefits. Sci., 2017. 356, 493–494. doi: 10.1126/science.aak9521.
22. Witoon, T., Bumrungsalee, S., Chareonpanich, M., Limtrakul, J., Effect of hierarchical meso macroporous alumina-supported copper catalyst for methanol synthesis from CO2 hydrogenation. Ener. Conver. Manag., 2015, 103, 886–894. doi: 10.1016/j.enconman.2015.
23. Borisut, P., Nuchitprasittichai, A., Methanol Production via CO2 Hydrogenation: Sensitivity Analysis and Simulation—Based Optimization. Front. Energy Res., 2019, 7, 81. doi: 10.3389/fenrg.2019.00081.
24. Xin, J-y., Cui, J.r., Niu, J.-z., Hua, S.-f., Xia, C.-g., Li, S.-b., Zhu, L.-m., Production of methanol from methane by methanotrophic bacteria. Biocatalys Biotransform., 2009, 22(3) 225-229. https://doi.org/10.1080/10242420412331283305
25. Singh, R., Ryu, J., Kim, S.W., Microbial consortia including methanotrophs: some benefits of living together. J. Microbiol., 2019, 57(11), 939-952. doi: 10.1007/s12275-019-9328-8
26. Ito, H., Yoshimori, K., Ishikawa, M., Hori, K., Kamachi, T., Switching Between Methanol Accumulation and Cell Growth by Expression Control of Methanol Dehydrogenase in Methylosinus trichosporium OB3b Mutant. *Front. Microbiol.*, 2021, 12, 639266. https://doi.org/10.3389/fmicb.2021.639266
27. Ma, X., Li, S., Ronda-Lloret, M., Chaudhary, R., Lin, L., van Rooij, G., Gallucci, F., Rothenberg, G., Raveendran Shiju N. & Hessel, V., Plasma Assisted Catalytic Conversion of CO2 and H2O Over Ni/Al2O3 in a DBD Reactor. Plasma Chem. Plasma Process., 2018, 39(1), 109-124. https://doi.org/10.1007/s11090-018-9931-1

28. Soldatov, S., Link, G., Silberer, L., Schmedt, C.M., Carbone, E., D'Isa, F.. Jelonnek, J., Dittmeyer, R., Navarrete, A., Time-Resolved Optical Emission Spectroscopy Reveals Nonequilibrium Conditions for CO 2 Splitting in Atmospheric Plasma Sustained with Ultrafast Microwave Pulsation, ACS Ener. Lett., 2020, 6(1), (124-130). 10.1021/acs energylett.0c01983,
29. Dobladez, . J.A.D., Maté, V.I.Á., Torrellas, S.Á., Larriba, M., Brea, P., Efficient recovery of syngas from dry methane reforming product by a dual pressure swing adsorption process, Internat J. Hydrogen Ener., 2021, 46(33), 17522-17533 10.1016/j.ijhydene.2020.02.153, (2020).
30. Wu, X., Lang, J., Jiang, Y., Lin, Y., Hu, Y.H., Thermo–Photo Catalysis for Methanol Synthesis from Syngas. ACS Sustain Chem. Eng., 2019, 7, 23, 19277–19285. https://doi.org/10.1021/acssuschemeng.9b05657.
31. Qin, Z., Zhai, G., Wu, X., Yu, Y., Zhang, Z., Carbon footprint evaluation of coal-to-methanol chain with the hierarchical attribution management and life cycle assessment. Energy Convers. Manag., 2016, 124, 168–179. https://doi.org/10.1016/j. enconman .2016.07.005
32. Khalafalla, S.S., Zahid, U., Jameel, A.G.A., Ahmed, U., Alenazey, F.S., Lee, C.-J., Conceptual Design Development of Coal-to-Methanol Process with Carbon Capture and Utilization. Energ., 2020, 13, 6421, doi:10.3390/en13236421
33. Meng, X., Wang, T., Liu, L., Ouyang, S., Li, P., Hu, H., Kako, T., Iwai, H., Tanaka A., Ye, J., Photothermal conversion of CO2 into CH4 with H2 over group VIII nanocatalysts: an alternative approach for solar fuel production. Angew. Chem., 2014, 126, 11662-11666. https://doi.org/10.1002/ange.201404953
34. Iglesias-Juez, A., Coronado, J.M., Light and Heat Joining Forces: Methanol from Photothermal CO2 Hydrogenation Chem., 2018, 4(7), 1490-1491. https://doi.org /10.1016/j.chempr.2018.06.015
35. Boukis, N., Diem, V., Habicht, W., Dinjus, E., Methanol Reforming in Supercritical Water. nd. Eng. Chem. Res., 2003, 42, 4, 728–735. https://doi.org/10.1021/ie020557i.

5

Microalgal Biomass Synthesized Biodiesel: A Viable Option to Conventional Fuel Energy in Biorefinery

Neha Bothra, P. Maniharika and Rajesh K. Srivastava*

Department of Biotechnology, GIT, Gandhi Institute of Technology and Management (GITAM) Deemed to be University, Rushikonda, Visakhapatnam, A.P., India

Abstract

Currently, fossil fuel sources (petroleum oils or gas and coal) provide 90% or more of total energy consumption that creates environmental issues in society and the world. So, developing alternative and pollution free fuel sources for the world that can help in more energy security and clean environment development for all the biotic components of the earth via the mitigation of toxic gases or particulate matter is needed. Microalgal diesel is reported as a renewable energy source (that comes in third generation biofuel) with reduction up to 80% of toxic or polluted gases or particulate matter. There are several sources of raw materials (cooking oil, vegetable oils, or many other sources of lipids) used for biodiesel synthesis that also utilize chemical processes (transesterification) for conversions and then use fuel energy sources for transportation tasks. Many other biofuels can be made up using chemical reactions, fermentation, and heat energy to breakdown starches, sugars, and other complex carbohydrates and biodiesel can be refined to produce fuel energy for daily energy requirements. Except for biodiesel, other biofuels (bioethanol or n-butanol) are frequently used as pure or blended with gasoline fuels for transportation tasks. Biodiesel can be derived from fats or lipid contents that are synthesized by green plants, including algal or microalgal species. Biodiesel can be derived from waste oils and can also be extracted from grasses, algae, animal wastes, cooking oil, wastewater sludge, or others. This chapter discusses several parameters for microalgal biomass cultivation and extraction and also transesterification processes that are used for biodiesel synthesis.

**Corresponding author*: rajeshksrivastava73@yahoo.co.in

Prakash Kumar Sarangi (ed.) Biorefinery Production of Fuels and Platform Chemicals, (105–130) © 2023 Scrivener Publishing LLC

Keywords: Biodiesel, microalgae, growth, parameters, lipids, biomasses, transesterification, conventional diesel

5.1 Introduction

Fuel energy is required for functions of domestic and industrial works and present and future periods are becoming the big challenges of fulfilling the ever increasing energy demand for the world and refinery synthesized diesel, petroleum, or any other fossil fuel is a non-renewable resource that can create global issues in the world. The major concern here is the rate of usage or consumption of energy fuel sources (many fold higher than the rate of replenishment) for our daily and industrial task operation by three to four decades. Many developed countries face shortages of fossil fuels storage or stocks and as of now, there is no alternative for these fuels. These fuels can cause a lot of pollution on the planet, which is one of the major problems faced by people and animals [1].

Biofuel has shown more potential to serve as a fuel source without disturbing the food supply. Traditional fuels are non-renewable energy sources and require thousands of years to synthesize this fuel energy, whereas biofuels are non-conventional energy sources with the least or non-polluting nature of the environment in the present and upcoming future. Fats or lipids can be extracted from vegetable oils, animal fats, and recycled cooking grease and can be blended with petroleum-based diesel. Sometimes, pure biodiesel can create problems in older vehicles in cold weather. These sources of energy can be used in transportation tasks (vehicles, flights and shipping) and other operations as required for daily life via the mitigation of greenhouse gas emissions at a global level [1, 2].

In recent years due to an ever increasing population rate in the world, the energy consumption rate has increased four to five-fold compared to ancient period energy needs and time developmental tasks. In this regard, the whole world has faced an energy crisis for the past many decades and it has forced the world to develop clean or renewable nature fuel sources with the least polluting nature to the environment [3]. In recent years, there has been increased concern to develop alternate sources of fuel production that do not cause any harm to the environment (without or least greenhouse effect). These alternate sources are found as biodegradable and non-toxic fuels and government agencies of different countries need to invest a lot of money in advanced research works that can help to develop alternative fuels to fossil fuels, one of the most accepted and promising alternatives being biofuels [4].

In algal plants, it has been reported that molecular events of the photosynthesis process are found to regulate the conversion of light energy into microalgal biomass and its products. These approaches and analysis have confirmed for targeting the enhanced productivity of microalgal biomass that has been seen to utilize or apply mutagenesis, strain selection, and genetic engineering. A number of reports have discussed their research outcomes, reaching success utilizing promising technology for achieving significant contributions to future microalgae based biotechnology approaches and processes [5]. In this regard, it has found adaptations for a wide range of environmental conditions for microalgal species cultivations which have evolved to greater genetic diversity within polyphyletic groups. These properties of algal biomass make rich sources of useful metabolites, including biodiesel that can be utilized in our society with a significantly lower demand for carbon emitted biofuel that can help our environment and energy security. These microalgal biomasses have shown potential for reducing the cost of biodiesels that can be used at a large scale for energy fuel purposes and, in this regard, it is still a challenge for the average productivity of most of microalgal biomasses that constraints for the industrial level [5, 6]. It showed a lower value of caloric energy than the maximal theoretical estimation. It is needed for identification of factors that limit the biomass yields and for removing the bottlenecks in lipid content storages. This effort can help as a provital of domestication strategies for making algal derived bioproducts profitable at an industrial scale [7].

In the current period, microalgal biofuels, including biodiesel, can help enhance energy fuels stocks for ever-increasing energy demand. It can also later satisfy for economic feasibility of overwhelming capital investments with its best operations. Microalgal plants can provide a high value of co-products with biodiesel formation and it can only be possible by doing the extraction of a fraction of microalgal biomasses [8]. This can improve the economies of local people via utilization of microalgal biorefinery benefits. There are many examples of high-value products from microalgal biomasses, such as pigments, protein, lipids, carbohydrates, vitamins, and antioxidants. These high-value products find their applications in the cosmetic, nutritional, and pharmaceutical industries [9]. Further, promotion of innovative microalgae biorefinery structures can be implemented for multiple product synthesis, including biodiesel, showing high values for society at a smaller price or cost. Many challenges like extraction strategies used for high-value products from microalgal biomasses can further be used in biorefinery development [8, 9].

In the modern period, microalgal biofuels at a large scale has faced many technical challenges, such as current growth and development of microalgal biomass for the biorefineries industry, and is still considered economically unviable. There is recent literature on upstream methodologies that are involved in microalgal-based biofuel development processes [7]. It needs a better understanding of the conversion of light energy into biomasses that are later used for biodiesel conversion tasks. Analysis of the productivities of most recent microalgal culture systems show that they are able to utilize wastewater sources as nutrient sources. Further, it should be noted that the co-location of the microalgal biomass production can be used at the industry level with innovative approaches for extraction and transesterification processes that can improve biodiesel synthesis [7, 9].

Nowadays, biofuels are regarded as the best alternative source via fulfilling the upcoming energy demand for world. It can help control the excess release of carbon dioxide from the fossil fuel sources, which is one of the major reasons behind the greenhouse effect at a global level, which causes melting of glaciers (can increase the level of sea water level) and climatic change effect [10]. This change can cause many natural disasters such as extreme rain or flood, drought, and others. So, the replacement of these fossil-based fuels with biofuels can eventually decrease the level of carbon dioxide and also help in maintaining the sea water level [11]. Among many biofuel sources in the world, biodiesel is found as the most sustainable source of energy and it can be produced from different feedstock's including algal plant biomasses that contain lipid or carbohydrate rich sources. Feedstocks can be any species of microbes or plant or any plant or animal products. Mainly, biodiesel has been produced from microalgae, yeast, waste vegetable oil, and non-edible oil (like Jatropa oil, Pongamia oil, Jojoba oil, Cottonseed oil, Linseed oil). Further, Deccan hump oil, Mahua oil, Kusum oil, Deccan hemp oil, Neem oil, Palm oil, etc. can also be utilized for biodiesel development. It can also be produced from waste and animal fats, as shown in Figure 5.1 [12].

Deccan hump or Mahua products (Neem, Jatropa, or palm oils) are shown to be the best substrates for the production of biodiesel and might end up in a competition for food versus fuel. Problems like food shortages can be solved in upcoming research works that can help to utilize other sources like many algal biomasses or their lipid contents and animal fats [13]. Vegetable oils, animal fats, or cooking oil, when collected, are subject to the transesterification process in which alcohol and fatty acid alkyl esters are formed by reacting glyceride with an alcohol (methanol or ethanol) [14, 15]. In this chapter, the author will emphasize the different parameters that affect biodiesel production (such as temperature, reaction time, hexane to oil molar ratio, type and amount of catalyst, mixing intensity, and

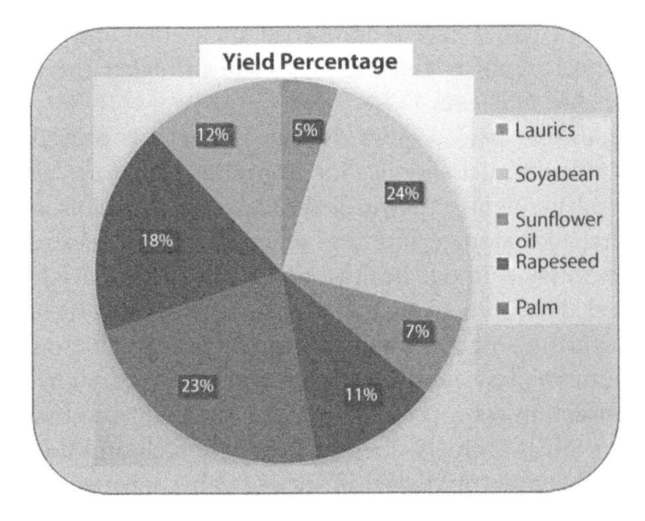

Figure 5.1 Various forms of oil or lipids (sources from higher plants) used for biodiesel production.

free fatty acid and water content). There are also different processes for extraction and the common process is reported as transesterification for biodiesel development [15].

5.2 Diesel

Refinery diesel is an aliphatic (hydro carbon) chemical product obtained during the chemical process of partial or fractional distillation of petroleum fuel and is also called petro-diesel, whereas its alternative, biodiesel, is a biosynthesized fuel source that uses plants photosynthesis processes for development of bio-molecules (lipids, protein, or carbohydrate sources) and is used as raw materials for biofuels or biodiesel development, which is completely different from petroleum derived diesels in terms of toxic gas emissions [16]. Demand of fuel energy is ever increasing globally due to luxury lifestyle and it needs more energy. In current periods, the main source of fuel energy is fulfilled by petroleum derived liquid fuels like petro-diesel and it contained more carbon chains in their chemical structure and showed greater energy content, as well as more competency in traditional engines for transportation tasks. It works as a traditional fuel source due to its high efficiency and the smaller cost that is preferred for all types of engine compared to other fuels, from the first diesel engine to present day cars. Diesel engines have found broad application and easy combustion in transport motor engines as it has shown higher thermodynamic efficiency, i.e., fuel efficiency [17].

From the ancient period onwards, petro-diesel has proved to be one of the most efficient fuel energies for transporting engines and now, we need to be independent of petro-diesel fuels that have lots of problems such as future stock storage and huge amounts of toxic gases and particulate matter emission in our healthy environments. We need to proceed with biodiesel development in the world without petro-diesel combustion in transporting engines. Biodiesel can be used right from today for industries, automobiles, and ships to agricultural sectors or anywhere with energy consumption tasks [18]. We can put forth effort to lower the cost of biodiesel or other biofuels compared to fossil fuels and these attempts can also help the underprivileged in society to live a better life and work efficiently in their respective field. It was discussed earlier that this efficient measure can come with many disadvantages. During combustion of petro-diesel fuel, it is reported to recombine with atmospheric oxygen and nitrogen and form nitrogen-oxides (NO_X) or sulfur-oxides (SOx) [19]. That can lead to the formation of smog or acid rain via affecting to biotic components, including human health and property. It releases huge quantities of carbon dioxide and carbon monoxide to our healthy environment that creates global problems for society. Table 5.1 shows the chemical compositions of these hazardous gases. This petro-diesel combustion or burning has increased the level of CO_2 that eventually increases the surrounding temperature, finally leading to global warming [18, 19].

Petro-diesel can exhaust from vehicles and is found to be harmful to human, animal, and plant growth and development in an unhealthy environment. This fuel has generated huge quantities of different sized PM and exposure to these particulate matters can lead to many pulmonary diseases in children or and the elderly. This fuel combustion can directly affect the CNS of animals [19, 20]. This type of fuel is a non-renewable source of energy which will be depleted eventually, so it is required to find any alternative source of energy which is sustainable to the environment, i.e., meets the needs of the present without compromising the ability of future generations to meet their own needs [20]. Problems related to energy crises can be solved by upcoming of renewable sources of energy, i.e., biodiesel development. It offers unlimited solutions to the issue of depleting reserves and harmful emissions of other fuels [21].

Table 5.1 Physical characteristics of petro-diesel [24, 25].

Properties	Values (unit)
Specific Gravity	0.88 kg/m^3
Kinematic Viscosity at 40°C	4.0 to 6.0 m^2/s
Cetane Number	47 to 65
Higher Heating Value	-1.28 $x10^6$ (Btu/gal)
Lower Heating Value	-1.19 $x10^6$(Btu/gal)
Density at 15.5°C	7.3 (lb./gal)
Carbon	77 wt.%
Hydrogen	12 wt.%
Oxygen	11 wt.%
Boiling Point	315-350 °C
Flash Point	100-170 °C
Sulfur	0.0 to 0.0015 wt.%
Cloud Point	-3 to 12 °C
Pour Point	-5 to 10 °C

According to the suggestions of many active researchers, biodiesel is needed to establish feasible fuels for today's traditional engine. Biodiesel can eradicate environmental issues like global warming and can help organisms live in a sustainable manner. It needs significant approaches in development for the domestic, industrial, and automobile economies [22]. Biodiesel will, more importantly, not have to invest huge amounts of money in its extraction process, unlike fossil fuels. We need to grow more oil plants, which again can be good for our environment. With upcoming advanced research and easing the technology, petro-diesel can be completely replaced with biodiesel that will also prove to be an efficient fuel [23].

5.2.1 Biodiesel

Biodiesel is a renewable, biodegradable fuel that can be produced from vegetable oils, animal fats, or recycled grease. It is found as liquid fuel (often referred as B100 or pure biodiesel) in unblended form. Biodiesel performance in cold weather can depend on the blending ratios of biodiesel with gasoline [24, 25]. Blends with a low percentage of biodiesel can work in better ways in cold temperatures. They are reported to contain some compounds that crystallized in very cold temperatures [26]. So, in winter periods, crystallization is overcome by adding a cold flow improver, so fuel with appropriate blend is required for the best performance in cold weather [20]. Biodiesel is generally produced from vegetable oils, grease, animal fats, etc. There are different processes for synthesizing biodiesel like blending, micro-emulsion processes, thermal cracking processes, and, most conventionally, transesterification processes. This is because this method is relatively easy and can take place at normal conditions [26, 27].

Biodiesel can blend in different concentrations like B5 (~up to 5% biodiesel), B20 (~6-20% biodiesel), or B100 (~ up to 100% biodiesel or pure biodiesel). B5 and B20 are found as lower blending ratios with gasoline fuels and these have shown a good balance of cost and least toxic gases or PM emission, as well as better cold-weather performance with material compatibility. Biodiesel has shown its ability to act as a solvent, whereas B100 are shown as high level blended ratios that require special handling and engine equipment modifications [17, 18]. Biofuels are produced by converting the waste or plant or animal synthesized fats and oils into biodiesel and glycerin (co-product) which are also produced via utilization of transesterification. This process takes place in the presence of a suitable and efficient catalyst (NaOH~ sodium hydroxide or KOH~ potassium hydroxide). Glycerin (a co-product) is a sugar rich compound that is used in manufacturing pharmaceutical products and cosmetics [28].

Biodiesel is generally produced in the transesterification process and is not used directly in vehicle fuels in combustion engines for much transportation. In earlier periods, biodiesel was made from vegetable oils, grease, and animal fats and advanced scientific or technological procedures or methods were applied to produce the potential biodiesel feedstock from many species of algae (micro or macro-plant cells structures). These plant biomasses can produce very high yields of lipids or carbohydrate rich products from a smaller area of non-fertilized lands or water bodies (fresh or marine nature). Biodiesel can be generally and easily transported through trucks, train, or barge sources. B5 nature of biodiesel is easily shipped through pipelines [29]. Biodiesel is a renewable and nonconventional energy source that can provide a substitute for current energy, i.e., petroleum diesel. Biodiesel is used as a vehicle fuel while increasing energy security in the world. This can improve air quality and the environment by providing safety benefits. Biodiesel has provided the positive energy balance to the environment and it can yield 4.56 units of fossil energy consumed over its life cycle [30]. Using biodiesel, a reduction in air pollution is reported due to a minimum level of emission of CO_2 (during combustion of biodiesel). Algal biomass cultivation is found to absorb the environmental CO_2 via growing as a feedstock that is used to produce the sustainable fuels. Biodiesel can cause less damage than the burning of petroleum diesel. In its spilled condition, it is found to be less combustible and safer for the environment. Overall, biodiesel is found to be safe for handling, storing, and transporting [31].

5.3 Production of Biodiesel

5.3.1 Origin of Biofuels

The production of various types of biofuels are divided into various generations based on their origin or development levels. First generation biofuels are synthesized from different types of food crop grains that are used as the main source of its feedstock and its products, biodiesel and bioethanol, are good examples that were mostly extracted or synthesized through microbial fermentation and transesterification process of oils, sugars, and starches that are available in various types of plant biomass [32]. Feedstock of first generation biofuels are carbohydrates of corn, wheat, and sugarcane, or lipid contents of rapeseed and soybean, etc. In this generation, since food crops are used as feedstock, the main disadvantage of using it is increased CO_2 footprint. While use of food crop originated biofuel can reduce greenhouse gases, the competition with global food demand might raise the fuel versus food debate. For dealing with these drawbacks, second generation biofuels were introduced that are synthesized from non-food products including waste organic matters or agri-residual waste (containing lignocellulosic contents) and can be seen in Figure 5.2 [33].

In second generation biofuels, the main feedstock is non-food crops and waste source plant biomass. Its best examples are switchgrass, waste vegetable oils, Jatropa, sugarcane bagasse, municipal solid wastes, and other lignocellulosic agricultural wastes. This biofuel generation has also faced some technical or compositional variation challenges. These are reported as different varieties of composition (need different processing requirements)

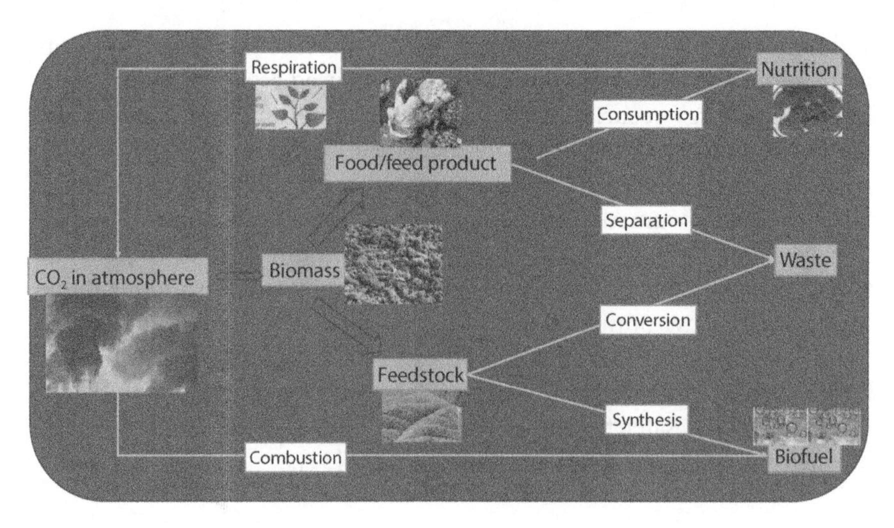

Figure 5.2 Plant biomass sources for energy & edible products & biomass synthesized from plants, including algae, that utilized photosynthesis.

and different varieties of plant products in different regions. So, it is difficult to choose an ideal feedstock for biofuel production that can be seen to vary in terms of performance, technology, scale, operating conditions, and target markets [33, 34]. Third generation biofuels are obtained from utilization of algal biomass which is a good feedstock for this generation of biofuel. It was reported that this biomass yield is ten times higher than the other lipid rich agricultural crops. Its main benefit is a large scale production of biomasses in a small area of land or cultivation sources. Algal species have shown higher photosynthetic efficiency in biomass production and is non-competitive with food crops, while also having a high lipid yield that makes it a more sustainable feedstock compared to the other two generations [34–36].

5.3.2 Biodiesel Production from Algae

Generally, algae biomasses are found in two types (i.e., macroalgae and microalgae) based on cellular nature. For biodiesel production, macroalgae is seaweed that can be used as alternative feedstock to produce biodiesel and is used in place of those depleting fuels that caused pollution also. Some of these macroalgal species are reported to contain more than 50% of lipid or oil content which can be extracted and also processed to use as transportation fuel [26]. Microalgae species have shown some different benefits such as a fast growth rate (requires less area for mass growth) and that it can be grown on wastewater in order not to compete with edible crops. It can be harvested daily and grown in all seasons. Its main advantage is the use of its waste for production of feed for many animals and many other purposes [25]. Both species of micro and macro algae can be cultivated in a short period of time at a large scale. The most used algal type (for biodiesel production) is microalgae, as it has shown the best photosynthetic and heterotrophic in nature, exhibiting the best potential to be grown as energy crops. They can produce high quantities of fats and oils which can be further processed to form biofuels or biodiesel. These biofuels or biodiesel are found to be ecofriendly in nature (i.e., do not produce any harmful gases after combustion) [37].

Various species of microalgae are microscopic in nature and these are found to cultivate in marine and freshwater environments. These are unicellular; they exist either individually or in groups. They have better photosynthesis ability and are also shown to use the greenhouse gas component (CO_2). They produce a huge quantity of O_2, helping in the maintenance of a healthy environment, are considered as photoautotrophic microbes, and are known to produce half of the atmospheric oxygen [38, 39]. Microalgae has been identified to be appropriate feedstock for production of biofuels (including biodiesel) due to their high lipid contents. It can in fulfilling the increasing energy demands of the world. Biofuel or biodiesel production processes are reported to comprise of two types of processes: upstream and downstream processes. This renewable energy can meet nearly 20% or more of global energy demand in coming periods. The main problem related to microalgae is the selection and improvement of its effective strain biomass production and the processing steps for biodiesel production [29, 34].

For replacing fossil fuels, this type of carbon source is the main objective for the best source of raw materials for biodiesel production and utilized biomass nature (simple or complex) can be decided for achieving the best yield of biodiesel, as shown in Figure 5.3. Biomasses are reported to contain all the elements that are found in fossil fuels. Main biomass production is found to involve the capturing of solar energy which fixes the carbon into storage bio-molecules (lipids or carbohydrates rich products) during photosynthesis processes [28, 29]. The key challenge of processing of microalgae biomasses is the dewatering process and it requires a high amount of energy as well as a significantly higher processing cost (accounts for 20-30% of the total cost) [29]. The dewatering process is divided into the first stage (thickening) and second stage (concentrating the biomass). This algal biomass is further processed through hydrothermal liquefaction (HTL) or transesterification processes to produce biocrude oils that

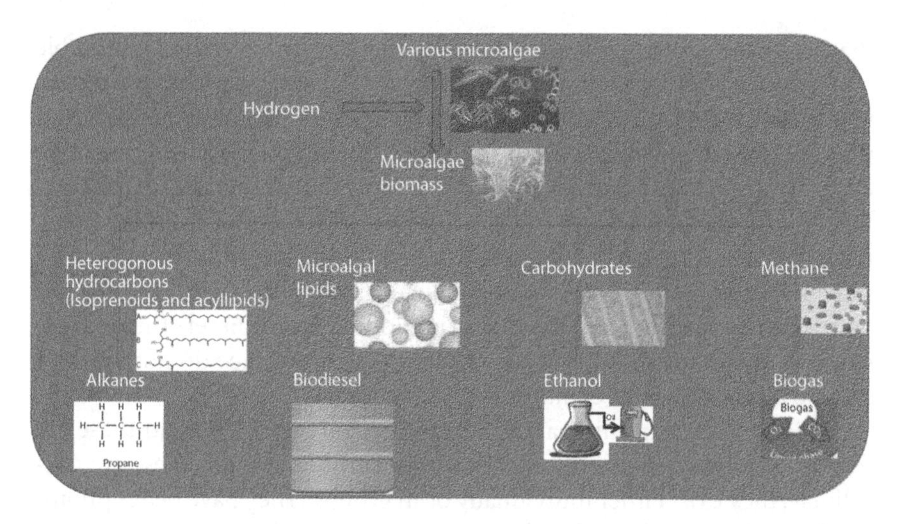

Figure 5.3 Microalgal species used for various types of biofuels including biodiesel.

are renewable diesel blend stock (RDB), fatty acid methyl ester (FAME), or biodiesel, respectively. After the extraction of these lipids, residuals can also be processed further to obtain biogas, ethanol, and biofertilizers [7, 45].

Chemical composition of algae is found in differing in growth phases and with an exponential growth or log phase of algae contains more protein. The stationary phase of algae contains a higher quantity of lipid, glycogen, and carbohydrate. This lipid content is found to vary from individual species or strains between or within taxonomic group [20]. Screening studies can help in suitable selection of efficient algae and the largest taxonomic group is green algae. These species of algae can be isolated easily from vast habitats and can be grown faster than other species or strains of other taxonomic groups [46]. There are many parameters which affect the growth of microalgae, including extraction and harvesting the lipid from the respective microalgae. Parameters affecting are as follows.

5.3.3 Intensity of Radiant Light

For algal biomass production, the primary energy source is radiant energy from sunlight that is captured during photosynthesis processes. There are mainly three types of microalgae that are found, based on nutritional source of cultivation. They are phototrophic, heterotrophic, and mixotrophic. The phototrophic nature of algal biomass cultivation is the conversion of atmospheric CO_2 in the presence of sunlight either in an open or closed system. The heterotrophic nature of algal biomass production has found very high operating costs due to the requirement of carbon sources as a supplement [40]. Mixotrophic algae are those species that have the ability to alter their mode of nutrition on the basis of availability of energy in the environment. Algae are grown in two physiological conditions: low intensity of light and high intensity of light. Low light adapted algal plants contained high chlorophyll content, whereas high light adapted algal plants contained low chlorophyll content. During cultivation, there is a mass production of algae that were found to be more optically dense with high irradiance. So, the growth rate can be reported to a limited concentration [35]. Limitation of algal growth rate is found due to limited penetration of light in the lower part of the culture due to either self-shading or absorbance of light by water. So, turbulence is normally used to avoid this type of condition, but this is not sufficient and not suitable to completely solve this problem. This situation can be improved by the photon conversion efficiency (PCE) method. The improving PCE method can increase the yield of biomass [41].

5.3.4 Lipid Content

In algal biomass cultivation, low lipid yield or accumulation is one of the major drawbacks for the commercialization of microalgae biodiesel development. Lipid levels in algal biomass are dependent on a variety of other factors such as photosynthesis performing ability, pH, CO_2 concentration, temperature, turbulence, inhibitors, and contamination. Temperature and light can have a direct effect on the growth rate of microalgae in any conditions (open or closed system). In short, a higher lipid content can produce higher yields of biodiesel. Similarly, lower lipid content can go for lower yields of algal biodiesel [47].

5.3.5 Biomass Culturing Conditions

Algal biomasses are reported to cultivate in mainly two types of system: open and closed system. These processes can also be done in either batches or continuously, depending upon the cultured species, pH, temperature, amount of algal biomass, etc. [42]. An open system is open raceway ponds (ORPs) and closed system is closed photobioreactors (PBRs) that are frequently used in cultivation of algal biomasses. They have shown their own sets of advantages and disadvantages during biomass cultivation. The selection process is totally dependent on the algal species that requires optimum conditions like temperature, pH, water content, etc.

ORPs are oval shaped systems with a depth of 1.0-3.0m with paddle wheels for water circulation. They have shown a high water demand for cooling of cultures through the evaporation process. Natural ORPs are ponds, deep sea, etc. ORPs are normally found to be cheaper with high efficiency techniques. It does not required any energy or manpower for treatment of wastewater sources that disturb the actual process, but it has reported some major drawbacks that include microalgal cells at the bottom that cannot utilize light properly with a high rate of evaporation, low mass transfer, and less diffusion of CO_2 into the atmosphere. These also required large areas for cultivation compared to the PBR system. These are reported as the most important drawbacks, along with contamination of water in the pond either biologically or chemically. Also, it is affected by the surrounding environment [43].

PBRs are reported as more controlled systems of biomasses and can be easily optimized on the basis of our requirements. Our requirements are found to depend on characteristic features of algal species (i.e., the condition for particular species shows the maximum growth rate). It has exhibited several advantages over open systems with several disadvantages. It has exhibited an absolute control over biomass culture conditions via preventing evaporation, reduced output losses, and also preventing contamination [43, 44]. This can, in turn, increase the yield and offer a protected and safer environment. With several advantages, PBRs also show many disadvantages such as increased concentration of O_2 that can reduce the photosynthetic ability, overheating, biofueling, high cost of building and operating biomass cultivation systems, cell damage by shear stress, and difficulty in scaling up of processes. It also required high energy consumption and mechanical agitation limits the use. There are several types of PBRs and the most used PBR is an airlift reactor. There are several factors which affect PBRs performance that indirectly affect the production of biodiesel [44].

5.3.5.1 *Temperature of Cultivation*

Temperature is the major factor responsible for microbial growth that can cause dense algal biomasses. As the temperature increases, the growth of biomass is increased up to a maximum level at the optimum temperature. Further, increase in temperature is reported to decrease in the growth rate of algal biomasses. Different strains or species of algae have shown different optimum temperatures (ranging from 15°C-26°C) [43].

5.3.5.2 pH of Cultivation

Most of the algal species have favored the neutral pH (except for marine algal species). Marine species are grown preferring a slightly alkaline pH (nearly 8-8.4). Some species (like *Spirulina platensis*) can grow tolerating up to 9 pH. As the concentration of CO_2 increases, the yield of biomass is also found to increase. But, with an increase in concentration of CO_2, pH falls down and very low pH value can decrease the yield. It is very important to maintain constant pH during cultivation and this can be attained by adding a constant buffer solution. Sometimes, a rise in pH is also reported as beneficial in term of inactivating the unwanted microbes in microalgal culture [36].

5.3.5.3 Duration Period of Light of Cultivation

Light is the main source of energy for phototrophic algae. It is necessary for them to maintain and it can help them produce carbon bio-molecule sources by utilizing CO_2 and water. Optimum light energy capturing is necessary for fixation of carbon dioxide by algal biomasses. It is also reported as the major limiting factor for biomass yield and its generation. Further, increasing the light intensity can decrease the cell growth rate due to the photo-inhibition effect [31].

5.3.5.4 Carbon Uptake of Cultivation

Microalgae plants have shown the same feature as higher plants exhibit, i.e., they can fix atmospheric carbon dioxide. The main carbon sources for them are atmospheric CO_2 and soluble carbonates ($NaHCO_3$, Na_2CO_3). Carbon uptake is needed to be very accurate. It is reported that increased carbon uptake can cause detrimental effects to microalgal cell growth. This is due to increased concentration of CO_2, which can create environmental stress that causes biological reduction in the capacity of algal cells [11].

5.3.5.5 Oxygen Generation in Cultivation

During photosynthesis, there is formation of oxygen by green plants, including microalgal biomass cultivation. The increment of oxygen level in liquid culture can cause a toxic effect and reduces photosynthetic efficiency, so oxygen needs to be removed regularly. Accumulation of oxygen can cause a very serious problem. This is not reported as a major issue for open systems of algal cultivation [41].

5.3.5.6 *Mixing Rates of Cultivation*

Uniform distribution of each and every substrate is important for optimal algal biomass cultivation. The low mixing rate can accumulate more biomass that can create an uneven distribution of nutrient and carbon dioxide levels with the settling of some biomasses. This can decrease the productivity of biomasses, whereas at a high mixing rate, it can cause damage to the algal biomass due to shear force. Mixing or agitating tasks in cultivation are reported to require high amounts of energy. Some methods of mixing are found that work, including mechanical stirring, gas injection, pumping, etc. [26].

5.3.5.7 *Nutrient Uptake of Cultivation*

With carbon substrate or sources, there are other main elements that are required during algal cell growth. After carbon sources, another important nutrient is nitrogen sources that are the main constituent of nucleic acid and proteins. *Chloromonas* sp. was cultivated with four nitrogen sources and these are urea, sodium nitrate, ammonium nitrate, and potassium nitrate with good yield. The yield of biomasses was maximum for urea substrates and is found to decrease for sodium nitrate, then ammonium nitrate, and least for potassium nitrate [21].

5.4 Harvesting of Microalgae

Harvesting of microalgae is reported by applying the extraction of biomass from the culture. The main harvesting methods are utilization of sedimentation, centrifugation, and filtration with flocculation step in addition [11, 48].

5.4.1 Extraction of Oil

There are several ways to generate fuel energy from algae biomasses. One of the main important methods is transesterification. Before this method, there was an important challenge for the extraction of this oil from biomass cells. There are several methods of extraction (either mechanically or chemically) reported for biodiesel production, as shown in Figure 5.4. The chemical extraction technique is known as the solvent extraction technique and can be favored more than the mechanical extraction technique in terms of more output. Chemicals used in the chemical mode extraction are hexane, benzene, or di-ethyl ether. Hexane is found as the most used chemical because it is less expensive and non-hazardous compared to other solvents [36].

For extracting the algal oil, it needs two things: first, suitable chemicals such as n-hexane, methanol, or NaOH (shown in Figure 5.5) and second, an algal sample that can either be collected from a pond or closed system. After collecting the algal samples, they are collected for spreading under the sun for drying and then, these dried samples are ground into fine powder and passed through different micron sieves to help extract different sizes of algal biomass. Now, hexane is mixed with the ground algae to extract oil in separate funnel. This mixture is untouched for 24 hrs for settling and formation of two layers. Now, algal oil is separated from algal biomass through the filtration technique and this extracted oil is now kept in a water bath for evaporating hexane. The oil yield (wt %) was the calculated [18, 49].

Extracted oil efficiency = Mass of oil extracted (in grams) x 100 / The total mass of dried algae

Different algal species have different percentages of lipid yield. This also has different parameters which affected the quantity and quality of extraction. They are as follows.

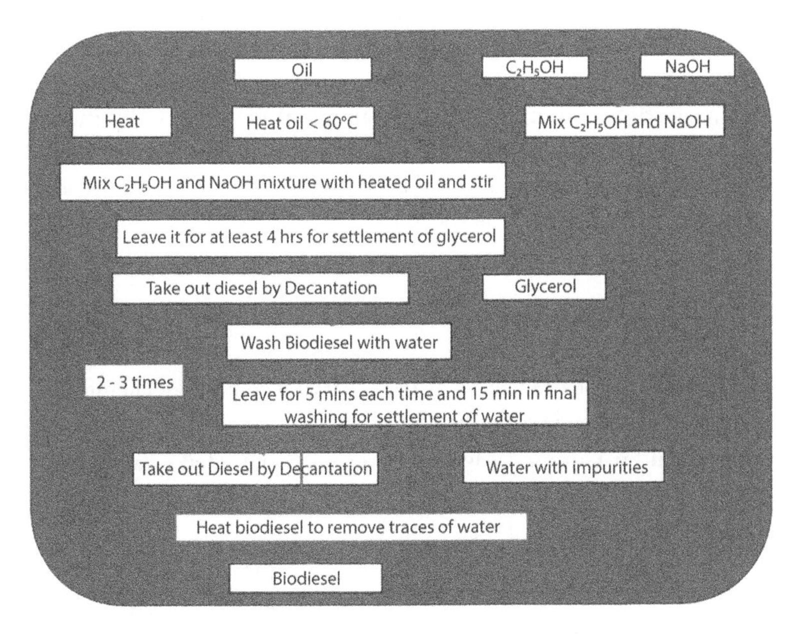

Figure 5.4 Schematic diagram for production of biodiesel from algal lipid or other lipids.

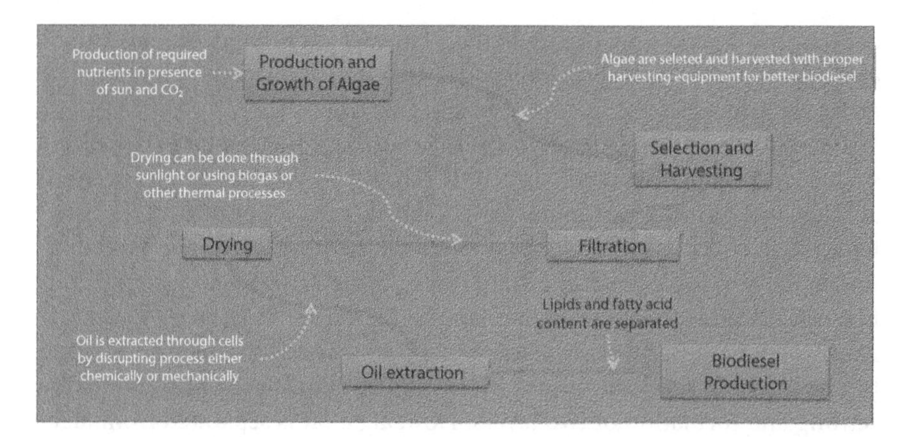

Figure 5.5 Transesterification processes for biodiesel synthesis used for conversion of lipids of algal or other sources.

5.4.1.1 Varying n-Hexane to Algae Ratio

By keeping algal biomass constant with the change of the volume of n-hexane that can change algae and n-hexane ratio, the percentage of yield of oil was found to increase as the solvent to algae ratio is increased. From the observation, it can be observed that the best optimum value for the biomass to n-hexane ratio is 1:2, as shown in Table 5.2a [28].

Many research scientists have believed in using these crops. It was observed that keeping a constant volume of n-hexane while varying the algal biomass shows an enhanced percentage of yield of oil by increasing the algae to solvent ratio. From this observation, it can be concluded that the yield is 2.5 times more than the equal ratio by increasing the algae to solvent ratio from 1:1 to 1:3, as shown in Table 5.2b. [30].

Table 5.2a Amount of extracted oil by using varying volumes of chemical solvent agents in biodiesel synthesis [28].

Sample of algal biomass (g)	n-Hexane (ml)	Algae to n-hexane ratio	Extracted oil (S)	Extracted efficiency %
1-30	30	1%	0.79	2.63
2-30	40	1.33%	0.92	3.07
3-30	50	1.66%	1.57	5.23

Table 5.2b Amount of oil extracted by using varying weights of dried algae in biodiesel [28, 29].

Sample with algal biomass (g)	n-Hexane (ml)	Algae to n-hexane ratio	Extracted oil (g)	Extracted efficiency %
1-20	30	1.5	0.7	4.1
2-15	30	2	0.8	5.33
3-10	30	3	0.82	7

5.4.1.2 *Varying the Algal Biomass Size*

Varying the size of algae can cause a change in the efficiency of oil extraction that can be done by sieving the algal biomass in different mesh numbers for getting different algal biomass sizes. It is observed that as the mesh size decreases (i.e., algal size decreases), there is an increase in extracted oil efficiency, as shown in Table 5.3a [34].

5.4.1.3 *Varying Contact Time between n-Hexane and Algae Biomass*

By varying the time of contact between n-hexane and algal biomass, there is a change in oil efficiency. Keeping both a constant volume of n-hexane and algal biomass and changing time of contact showed a change in efficiency of oil extraction. It was observed that as the time of contact increases, there is increased efficiency of oil due more interaction between biomass and solvent after some time, as shown in Table 5.3b [17, 18].

Table 5.3a Amount of oil extracted by using different sizes of algal biomass [34].

Sample of algae biomass (g)	Mesh number	Algal biomass size	n-Hexane (ml)	Extracted oil (g)	Extracted efficiency %
1-30	20	0.S41	50	1.55	5.16
2-30	30	0.595	50	2.02	6.73
3-30	50	0.297	50	2.3	7.66

Table 5.3b Oil extracted by using different contact time with solvents and algal biomass [17, 18].

Sample with algae biomass (g)	n-Hexane (ml)	Time (hours)	Oil extracted (g)	Extracted Efficieucy %
1-30	50	10	1.02	3.4
2-30	50	15	1.04	3.46
3-30	50	20	1.57	5.23

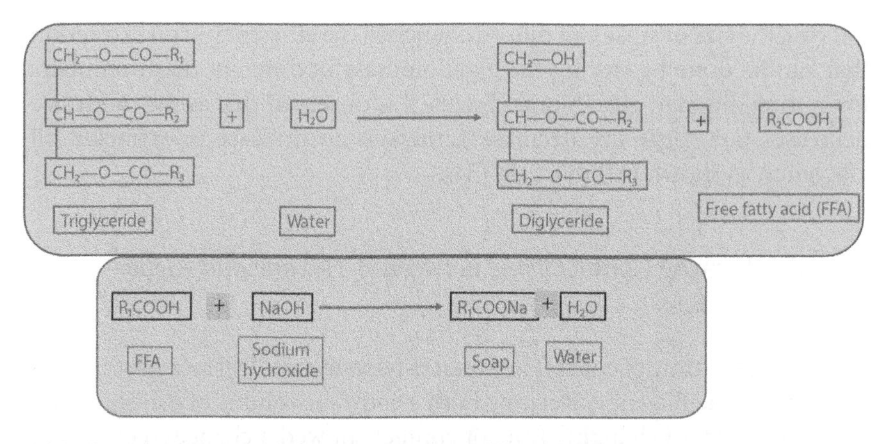

(**a**) Triglycerides Reaction with Water Diglycerides and Free Fatty Acids

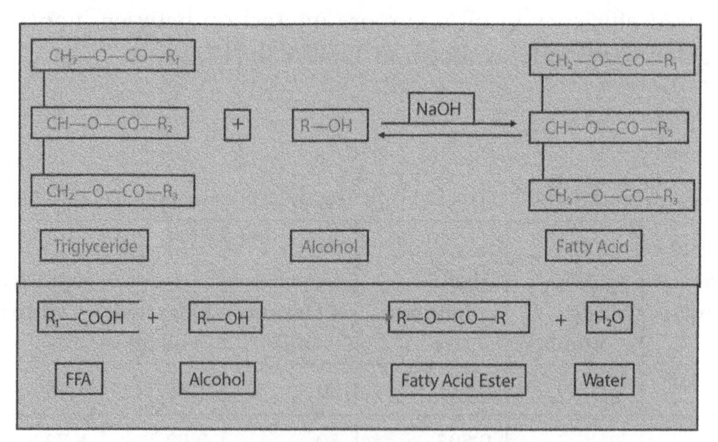

(**b**) Triglycerides Reaction with Free Fatty Acids showed Fatty Acids Synthesis

Figure 5.6 Chemical reactions occurred in transesterification reaction for biodiesel synthesis.

5.4.2 Transesterification

After extraction, algal oil extracted is converted into the product biodiesel. This can be done by various methods. The most commonly used method is transesterification. This method is also called an ester exchange reaction of microalgal lipids. It is a multiple step process and in this process, triglyceride is allowed to react with methanol in the presence of a catalyst. It can be catalyzed by either acid or base with methanol to get the corresponding fatty acid, but there is a disadvantage with acid, i.e., its corrosive nature. So, mostly alkali is preferred for this process, as shown in Figure 5.6 [17, 50].

5.5 Conclusion

In recent years, we have needed to use biodiesel from algal biomass at a commercialization scale and people have to develop the habit to utilize it for all activity or operations for country development tasks. Various types of other biofuels (ethanol, n-butanol, algal biodiesel or syngas, biogases, biohydrogen or methane) have been developed in the past few decades and it is needed to create awareness for utilization of biofuels including biodiesel from algal biomass or other lipids sources for domestic, industrial, and transportation functions. In this chapter, the author has discussed the different growth factors and bioprocesses for developing or synthesizing biodiesel and in this regard, algal species (macro or micro nature) have been cultivated in open (ORC) and closed systems (PBRs). Currently, people are dependent on utilizing conventional energy sources like petroleum and coal. This energy utilization has created lots of environmental issues that are affected by biotic components (including human or plants), as well as property health. Biodiesel synthesis was found to be affected by several parameters such as production or cultivation conditions (pH, temperature, intensity of light and its duration), appropriate microalgal species (storage of lipid or carbohydrates capability), extraction technology or procedure and transesterfication. All these factors are reported to affect the final quality of biodiesel with least or non-toxic gases or particulate matter emission that can help to maintain a healthy environment.

Abbreviations

CNS: Central Nervous System; CO_2: Carbon Dioxide; FAME: Fatty Acid Methyl Ester; HTL: Hydrothermal Liquefaction; KOH: Potassium

Hydroxide; Na_2CO_3: Disodium Carbonates; $NaHCO_3$, Sodium Hydrogen Carbonates; NaOH: Sodium Hydroxide; NO_X: Nitrogen-Oxides; ORPs: Open Raceway Ponds; PBRs: Photobioreactors; PCE: Photon Conversion Efficiency; PM; Particulate Matter; RDB Renewable Diesel Blend Stock; SOx: Sulfur-Oxides

References

1. Brennan, L., Owende, P., Biofuels from microalgae-A review of technologies for production, processing, and extractions of biofuels and co-products. Renew. Sustain. Ener. Rev. 14(2), 557-577, 2010.
2. Tiwari A., Kiran, T., Biofuels from Microalgae. Advances in Biofuels and Bioenergy, Nageswara-Rao M., Soneji, J. R., (eds.), Intechopen Limited London, EC3R 6AF, UK, 2018. pp-1-8
3. Abo, B.O., Odey, E.A., Bakayoko, M., Kalakodio, L., Microalgae to biofuels production: a review on cultivation, application and renewable energy. Rev Environ Health., 34(1). 91–99, 2019.
4. Refaat, A.A.; Different techniques for the production of biodiesel from waste vegetable oil. Int. J. Environ. Sci. Technol., 7, 183–213, 2010.
5. Benedetti, M., Vecchi, V., Barera, S. Dall'Osto L., Biomass from microalgae: the potential of domestication towards sustainable biofactories. Microb Cell Fact., 17, 173 (2018). https://doi.org/10.1186/s12934-018-1019-3
6. Caporgno M.P., Taleb, A., Olkiewicz. M., Font, J., Pruvost, J., Legrand, J., Bengoaa C., Microalgae cultivation in urban wastewater: nutrient removal and biomass production for biodiesel and methane. Algal Res., 10, 232–9, 2015. https://doi.org/10.1016/j. algal. 2015 .05.011
7. Chew, K.W., Yap, J.Y,, Show, P.L., Suan, N.H., Juan, J.C., Ling, T.C., Leeg, D.-J., Changhi, J.-S., Microalgae biorefinery: high value products perspectives. Bioresour Technol., 229, 53–62, 2017. https://doi.org/10.1016/j. biortech.2017.01.006.
8. Moreno-Garcia, L., Adjallé, K., Barnabé, S., Raghavan, G.S.V., Microalgae biomass production for a biorefinery system: Recent advances and the way towards sustainability. *Renew. Sustain Ener Rev.*, 76, 493-506, 2017, https://doi.org/10.1016/j.rser.2017.03.024
9. Brasil, B.S.A.F., Silva, F.C.P., Siqueira, F.G., Microalgae biorefineries: The Brazilian scenario in perspective. *New Biotechnol.*, 39, 90-98, 2017, https://doi.org/10.1016/j. nbt.2016.04.007
10. Marchetti, J.M., Miguel, V.U., Errazu, A.F., Possible methods for biodiesel production. *Renew. Sustain. Ener. Rev.*, 11(6), 1300-1311, 2007.
11. Folaranmi, J.; Production of biodiesel (B100) from Jatropa Oil using sodium hydroxide as a catalyst. *J. petroleum eng.*, 2013, 956479, 6, 2013.

12. Elkady, M. F., Zaatout, A., Balbaa, O., Production of biodiesel from waste vegetable oil via K.M. Micromixer. *J chemist.*, 2015, ID 630168, 9, 2015.
13. Gnanaprakasam, A., Sivakumar, V.M., Surendhar, A., Thirumarimurugan, M., Kannadasan, T; Recent Strategy of Biodiesel Production from Waste Cooking Oil and Process Influencing Parameters: A Review. *J. Enger.*, 2013, 926392,10, 2013.
14. Galadima, A., Muraza, O., Biodiesel production from algae by using heterogeneous catalyst: A critical review. *Energ.*, 78, 72-83, 2014,
15. Mondal, M., Goswami, S., Ghosh, A., Oinam, G., Tiwari, O. N., Das, P., Gayen, K., Mandal, M. K., Halder, G.N., Production of biodiesel from microalgae through biological carbon capture: a review. *3 Biotech.*, 7(2): 99, 2017.
16. Akubude, V.C., Nwaigw, K.N., Dintwa, E., Production of biodiesel from microalgae via nanocatalyzed transesterification process: A review. *Mater. Sci. Ener. Technol.* 2(2), 216-225, 2019.
17. Arous F., Atitallah, I.B., Mechichi, T., A sustainable use of low cost raw substrates for biodiesel production by the Oleaginous yeast *Wickerhamomyces anomalus. 3 Biotech.,* 7(4), 268, 2017.
18. Thangavelu, K., Sundararaju, P., Srinivasan, N., Muniraj I., Uthandi S., Simultaneous lipid production for biodiesel feedstock and decontamination of sago processing wastewater using *Candida tropicalis* ASY2. *Biotechnol. Biofuels.*, 13, 35, 2020.
19. Mohammed, A.T., Gashaw, A., Getachew, T., A review on biodiesel production as an alternative fuel. J Forest product. Indust., 4(2),80-85, 2015. Khan, S., Siddique, R., Sajjad, W., Nabi, G., Hayat, K.M., Duan, P., Yao, L., Biodiesel production from algae to overcome energy crisis. *HAYATI J.Biosci.*, 24(4),163-167, 2017.
20. Nautiyal, P., Subramanian, K.A., Dastidar, M.G., Recent Advancements in the production of biodiesel from algae :A review . Reference Module in Earth Systems and Environmental Sciences. Elias, S., Alderton, D.H.M. *et al.*, (eds), Elsevier Inc. pp. 1-12, 2014.
21. Yew, G.Y., Lee, S.Y., Show, P.L., Tao, Y., Law, C.L., Nguyen, T.T. C., Chang, J.S.; Recent advances in algae biodiesel production: From upstream cultivation to downstream processing. *Bioresour. Technol.Rep.*, 7, 100227, 2019.
22. De, A., Box, S.S., Application of Cu impregnated TiO as a heterogeneous nanocatalyst for the production of biodiesel from palm oil. *Fuel.*, 265,117019, 2020.
23. Ihsanullah, Shah S., Ayaz M., Ahmed I., Ali M., Ahmad, N., Ahmad. I., Production of biodiesel from algae. *J. Pure Appl. Microbiol.*, 9(1),79-85, 2015.
24. Baig, R.U., Malik A., Ali, K., Arif, S., Hussain, S., Mehmood, M., Sami, K., Mengal, AN., Khan, M.N., Extraction of oil from algae for biodiesel production, from Quetta, Pakistan. IOP Conf. Series: *Materials Sci. Eng.*, 414,012022, 2018.

25. Sharma, P,K., Saharia M., Srivstava, R., Kumar, S., Sahool L., Tailoring Microalgae for Efficient Biofuel Production. Front. Mar. Sci., 5, 382. 2018
26. Hossain, N., Mahlia, T.M.I., Saidur, R., Latest development in microalgae-biofuel production with nanoadditives. *Biotechnol.Biofuel.*, 12,125, 2019.
27. Martin, M., Grossman, I.E., Design of an optimal process for enhanced production of bioethanol and biodiesel from algae oil via glycerol fermentation. *Appl. Energ.*, 135,108-114, 2014,
28. Musa, M., Ayoko, G.A., Ward, A., Rösch C., Brown, R.J., Rainey, T.J., Factors Affecting Microalgae Production for Biofuel and the Potentials of Chemometric Methods in Assessing and Optimizing Productivity. *Cells*, 8, 8512019, 2019.
29. Medipally, S.R., Yusoff F.M., Banerjee, S., Shariff, M., Microalgae as Sustainable Renewable Energy Feedstock for Biofuel Production. Renewable Energy and Alternative Fuel Technologies. *Biomed. Res. Internat.*, 2015, 519513, 13, 2015.
30. Dickinson, S., Mientus, M., Frey, D., Amini-Hajibashi, A., Ozturk, S., Shaikh, F., Sengupta D., El-Halwagi, M.M., A review of biodiesel production from microalgae. *Clean Techn Environ Polic.*, 19, 637–668, 2017.
31. Chen, C.Y., Yeh, K.L,. Aisyah, R., Lee, D.J., Chang, J.S., Cultivation, photobioreactor design and harvesting of microalgae for biodiesel production: a critical review. *Bioresour. Technol.*, 102, 71–81, 2011.
32. Cuellar-Bermudez, S.P., Aguilar-Hernandez, I., Cardenas-Chavez, D.L., Ornelas-Soto, N., Romero-Ogawa, M.A., Parra-Saldivar, R., Extraction and purification of high-value metabolites from microalgae: essential lipids, astaxanthin and phycobiliproteins. *Microb Biotechnol* 8,190–209, 2015.
33. Freire, I., Cortina-Burgueño, A., Grille, P., Arizcun, M.A., Abellán, E., Segura, M., Sousa, F.W., Otero. A ., Nannochloropsis limnetica: a freshwater microalga for marine aquaculture. *Aquacult.*, 459, 124–130, 2016
34. Yew G.Y., Lee S.Y., Show P.L., Tao Y., Law C.L., Nguyen, T.T.C., Chang J., Recent advances in algae biodiesel production: From upstream cultivation to downstream processing. *Bioresour. Technol Rep.*, 7, 100227, 2019
35. Batista, F.R.M., Lucchesi, K.W., Carareto, N.D.D., Costa M.C.D., Meirelle A.J.A.; Properties of microalgae oil from the species *Chlorella protothecoides* and its ethylic biodiesel. *Braz. J. Chem. Eng.*, 35(4), 1383-1394, 2018
36. Islam A.M., Magnusson, M., Brown, R.J., Ayoko, G.A., Nabi, M.N., Heimann, K., Microalgal species selection for biodiesel production based on fuel properties derived from fatty acid profiles. *Energi.*, 6, 5676-5702, 2013.
37. Hossain N.,, Mahlia, T.M.I., Saidur, R.; Latest development in microalgae-biofuel production with nano-additives. *Biotechnol Biofuel.*, 12, 125, 2019.
38. Nizami, A., Rehan, M.; Towards nanotechnology-based biofuel industry. *Biofuel Res J.*, 18, 798–9. 2018.

39. Laurens, L.M.L., Chen-Glasser. M., McMillan, J.D.; A perspective on renewable bioenergy from photosynthetic algae as feedstock for biofuels and bioproducts. *Algal Res.*, 24, 261–4, 2017.
40. Goh, B.H.H., Ong, H.C., Cheah, M.Y., Chen, W.-H., Yu, K.L., Mahlia, T.M.I.; Sustainability of direct biodiesel synthesis from microalgae biomass: a critical review. *Renew Sust Energy Rev.*, 107, 59–74, 2019.
41. Schenk, P.M., Thomas-Hall, S.R,, Stephens, E., Marx, U.C., Mussgnug, J.H., Posten, C., Second generation biofuels: high-efficiency microalgae for biodiesel production. *Bioenergy Res.*, 1, 20–43, 2008.
42. Hasannuddin, A.K., Yahya, W/J., Sarah, S., Ithnin, A.M., Syahrullail, S., Sidik, N.A.C., Kassim, K.A. A, Ahmad, Y., Hirofumi, N., Ahmad, M.A., Sugeng, D.A., Zuber, M.A., Ramlan, N.A., Nano-additives incorporated water in diesel emulsion fuel: fuel properties, performance and emission characteristics assessment. *Energy Convers Manag.*, 169, 291–314, 2018.
43. Safarik, I., Prochazkova, G., Pospiskova. K., Branyik, T., Magnetically modified microalgae and their applications. *Crit Rev Biotechnol.*, 36:931–41, 2016.
44. Byreddy, A.R., Gupta, A., Barrow, C.J., Puri, M., Comparison of cell disruption methods for improving lipid extraction from thraustochytrid strains. *Mar Drugs.* ,2015;13:5111–27.
45. El-Sheekh, M., Abomohra, A.EF., Biodiesel Production from microalgae. In book: Industrial Microbiology: Microbes in Action Chapter: 16, Publisher: Nova Science Publishersm Garg, N, Aeron, A.; (eds), pp.1-13, 2016.
46. Pradana, Y.S., Sudibyo, H., Suyono, E.A., Indarto, Budiman, A., Oil algae extraction of selected Microalgae species grown in monoculture and mixed cultures for biodiesel production. *Energ. Proced.*, 105, 277-282, 2017.
47. Valero, E., Alvarez, X., Cancel, A., Sanchez, A., Harvesting green algae from eutrophic reservoir by electroflocculation and post use for biodiesel production. Bioresour *Technol.*, ,187,255-262, 2015.
48. Karmakar, R., Rajor ,A., Kundu K., Kumar N., Production of biodiesel from unused algal biomass in Punjab, India. *Petroleum Sci.*, 15(1), 164–175, 2018.
49. Mitrea, L., Ranga, F., Fetea, F., Dulf, F.V., Rusu, A., Trif , M., Vodnar, D.C., Biodiesel-Derived Glycerol Obtained from Renewable Biomass—A Suitable Substrate for the Growth of *Candida zeylanoides* Yeast Strain ATCC 20367. *Microorganis.*, 7(8), 265, 2019.
50. Cao, H., Zhang, Z., Wu, X., Miao X., Direct Biodiesel Production from Wet Microalgae Biomass of *Chlorella pyrenoidos* a through in situ transesterification. *BioMed Resear Internat.*, 2013, 930686, 6, 2013.

6

Algae Biofuel Production Techniques: Recent Advancements

Trinath Biswal[1], Krushna Prasad Shadangi[2*] and Prakash Kumar Sarangi[3]

[1]Department of Chemistry, Veer Surendra Sai University of Technology, Burla, Odisha, India
[2]Department of Chemical Engineering, Veer Surendra Sai University of Technology, Burla, Odisha, India
[3]College of Agriculture, Central Agricultural University, Imphal, Manipur, India

Abstract
This study describes the use of algae as a significant source of biofuel. Microalgae are considered as an unbelievable source of biofuel. For the production of 3rd generation biofuel, algae are subjected to a number of stages and its efficiency of performance in I/C engines is the key technology of the production of biofuel from algae. Algae biofuel can be produced using thermochemical and biochemical methods. Both the methods have some benefits along with the drawbacks. Direct combustion, thermochemical liquefaction, gasification, and pyrolysis are techniques to produce solid, liquid, and gaseous fuels. Biochemical methods of conversion yield biogas, bio-hydrogen, and bioethanol as the products. Apart from this, the extracts of the algae can be used to produce biodiesel. In this chapter a detailed description of the methods of conversion of algae biomass including Genetic Engineering toward biofuels production are given.

***Keywords*:** Microalgae, biofuel, pyrolysis, gasification, biodiesel, direct combustion

6.1 Introduction

Biofuel is the alternative to fossil fuels derived from biomass. The types of biofuel vary with the sources. The fuel produced from algae biomass

**Corresponding author*: krushnanit@gmail.com; kpshadangi_chemical@vssut.ac.in

Prakash Kumar Sarangi (ed.) Biorefinery Production of Fuels and Platform Chemicals, (131–146) © 2023 Scrivener Publishing LLC

is termed as algae biofuel. Fuel production from algae does not affect the food chain since algae is not a food supplement for human beings. Varieties of algae grow in the environment with the help of the nutrients available and sun light. As algae growth does not require any specific land, soil, or atmospheric conditions, cultivation is easy. It can grow in any climate condition, hence it is one of the good renewable sources to be used as a feedstock for the production of biofuel. Since the process of production is not economic and well developed yet, such process of production of fuel is not commercially implemented. The growth of microalgae needs sunlight, water, minerals, and CO_2 for photosynthetic activity. Temperature is one of the most important parameters for the suitable growth of microalgae, which is almost 20 °C to 30 °C. For decreasing the cost of production of the biofuel, we have to depend on freely accessible sunlight. There are various ways in which the biomass product of microalgae can be transformed into sources of energy. However, rigorous research is still being conducted to get the best way to produce the algae biomass. Algae biomass contains huge amounts of lipid content and is the major element used for biofuel production. Biofuel from algae can be produced through thermochemical and biochemical conversion techniques [1]. Every conversion techniques has their own advantages and draw backs. Thermochemical methods of conversion techniques convert the algae biomass to solid, liquid, and gaseous biofuels. Such methods include gasification and pyrolysis. However, the choice of conversion techniques is chosen based on the yield of a particular product, since it depends on various operating parameters. Pyrolytic oil can be used as a fuel or a platform for the production of chemicals. The yield is also a function of the physio-chemical composition of the respective biomass. Gasification is the second thermochemical conversion technique whose product is mainly gas. The gas is also called syngas, having a composition of CH4, CO, H2, and CO2. Such composition of gas again varies with the biomass source along with the process conditions and the types of gasifier. Moreover, the temperature, moisture content, the heat dissipation, and the elemental composition plays an important role for the quality of syngas produced. The syngas can also be converted to ethanol and methanol as chemical products in the presence of catalyst and at a particular temperature. Such chemicals can also be produced by biochemical methods of conversion that can be a replacement of fossil fuels and as a source for the production of various chemicals. The lipid content of the algae can be separated using solvent extraction techniques, which with a transesterification reaction, is converted to biodiesel. The biodiesel can be blended with conventional diesel to enhance the performance of the engine. Hence, algae can be a good source for the production of fuel as future energy. Algae can also consume CO released by the burning of such fossil fuels during their vegetation, hence it can be said that the biofuel is a carbon neutral fuel.

In this chapter the various methods used for biofuel production from algae are discussed along with their process parameters.

6.2 Technologies for Conversion if Algal Biofuels

6.2.1 Thermochemical Conversion of Microalgae Biomass into Biofuel

Thermochemical conversion means the breakdown of organic materials by the action of heat to form biofuel. It is generally carried out either by direct combustion, thermochemical liquefaction, gasification, or pyrolysis.

6.2.1.1 Gasification

The process of gasification is the incomplete combustion or oxidation of biomass into the flammable gas mixture at an elevated temperature (800–1000 °C). In the process of gasification, biomass undergoes a chemical reaction with steam and O_2 and form syngas, which is a mixture of CO, N_2, CO_2, H_2, and CH_4. The major benefits of gasification are the biomass-to-energy conversion path, which synthesizes syngas from diversified categories of significant feedstock. Syngas possesses low calorific gas usually 4 to 6 MJ/m^3 and on direct burring is used as fuel for gas turbines or gas engines. It was identified that the gasification of algae biomass at 1000°C produces 0.64g CH_3OH from 1g of biomass. The energy balance (ratio of CH_3OH formed to the total energy used) is 1:1, which indicates a positive energy balance of gasification. The low value of the energy balance is being recognized due to the use of an energy intensive centrifuge method during the yield of the biomass. It was found that the gasification of microalgae C. vulgaris along with nitrogen cycling produces biofuel enriched with CH_4 and all the components containing N_2 can be converted into nitrogenous fertilizer having quality similar to NH_3. This method of gasification for production of biofuel by microalgae biomass needs extensive research, particularly in the energy balance process of drying the microalgae biomass for gasification [2, 3].

6.2.1.2 *Thermochemical Liquefaction*

In this method, the wet microalgae biomass substances are converted into liquid biofuel. This process is carried out at high pressure (5–20 MPa) and low temperature (300–350°C) assisted by a catalyst in the presence of H_2 yield bio-oil. The reactor used for this conversion process is costly and the fuel-feed systems are usually complex. The major benefit of this process is that it is able to convert wet biomass material into energy. In the thermochemical liquefaction method, a greater quantity of water is activated in the sub-critical environment for degradation of microalgae biomass by the utilization of higher energy density. The microalgae B. braunii on thermochemical liquefaction at a temperature of 300 °C attains the highest yield of 64% dry wt. basis of oil with a higher heating value (HHV) of about 45.9 MJ/kg and is also identified as a positive energy balance of this method having an output/input ratio of 6.67:1 [4]. Similarly, in another study it was found that Dunaliella tertiolecta provides an HHV value of 34.9 MJ/kg with an energy balance of 2.94:1 [5]. Hence, by taking all these facts, it can be said that thermochemical liquefaction is considered as a feasible solution for conversion of microalgae biomass to liquid fuel.

6.2.1.3 *Pyrolysis*

Pyrolysis is the thermal decomposition of the microalgae biomass material at elevated temperatures (400–1000 °C) without oxygen and catalyst producing solid fuel (coke and biochar), liquid fuel (bio-oil), and gaseous fuel (CH_4 and some higher hydrocarbons in gaseous form). The method of pyrolysis is of two kinds:

- Slow pyrolysis with a slow rate of heating (0.1–1 °C/s), low processing temperature, and longer time duration
- Fast pyrolysis with a high rate of heating (>1000 °C/s), higher processing temperature, and shorter time duration

Normally, bio-oil is formed in the process of pyrolysis. Because of a higher percentage of ash present in algae, pyrolysis is treated as the most favoured conversion method, but there are some problems associated with it such as acidity, stability, and viscosity of the synthesized bio-oil. The aqueous phase contains low molecular weight compounds such as primary alcohol, ketone, and acid and the non-aqueous phase of oxygen comprises of substances like aliphatic compounds, carbonyls, alcohols, acids cresols, phenols, etc. along with aromatic hydrocarbons.

An experiment was carried out using pyrolyzed soyabean oil and pure soyabean and found that the pour point and viscosity of bio-oil obtained in the process of pyrolysis was much less. However, no noticeable difference in the heating value between the pyrolysis oils and raw soyabean is observed [6, 7]. The pyrolysis assisted by microwave irradiation is found in a rapid, effective process for the production of bio-oil. The pyrolysis of microalgae produces a lower hydrocarbon range of syngas, liquids, biochar, and bio-oil. It was reported that 44% of bio-oil was produced in the pyrolysis of microalgae biomass from the effluents of wastewater. The biomass obtained from an important macroalgae, Nannochloropis gaditana, produced bio-oil having the highest calorific value (12.6 MJ/Kg). In this method, the bio-oil obtained contains the majority of gaseous substances CH_4 and liquid materials $C_{10}H_{22}$. About 78% of bio-oil exhibits excellent results in the flash pyrolysis processing of microalgae [8]. The algae grown in an open pond system on pyrolysis at a temperature of 500 °C provides 59% of bio-oil having a heating value of 21 MJ/Kg. The process of slow pyrolysis is always less efficient and produces less concentration of biofuel in comparison to the fast pyrolysis process. The pyrolysis carried out at elevated temperatures, particularly, is less effective for the conversion to get the required fuel. Hence, a different strain of algae and the operating temperature at which the process of pyrolysis occurs plays a key role [9]. Catalytic hydrocracking is another significant method in which the biomass having high molecular weight are degraded into simpler materials having a low molecular weight by application of stream and hydrogen in the presence of an appropriate catalyst in the temperature range of 200–450°C. The methods of hydrogenation and cracking both are complementary to each other. The biomass cracking produces lower olefins by absorption of heat energy and is an endothermic process, whereas the process of hydrogenation is an exothermic process and produces heat energy due to cracking. The algal oil extracted after processing gets converted into beneficial biofuels containing an appropriate range of hydrocarbons. This method of conversion is generally carried out by either a heterogeneous or homogeneous catalytic system [10]. The homogeneous catalysts applied in the process of transesterification is the cause of cracking the oil, but the decrease in pH with an increase in acidity of the medium can be a cause of corrosion of the machinery parts. A microporous catalyst HZSM-5 having a haphazard channel can synthesize bio-oil of 52.7 wt.% with the microalgae biomass. The aromatic compounds present in the biofuel are of about 26 wt.%. The biofuel extracted from HTL (hydro thermal liquefaction) of the algae biomass product in combination with HVGO (Heavy vehicle gas oil) retards the rate of conversion, but increases the catalytic activity of cooking. Again, it limits the economic appropriateness of the algal bio-oil on blending with HVGO. The process of pyrolysis increases the yield % of oil from the microalgae Chlorella prothothecoides by altering its metabolic path in the direction of the heterotrophic growth [11].

6.2.1.4 Direct Combustion

In the method of direct combustion, the microalgae biomass undergoes combustion in the presence of oxygen or air and converts the chemical energy stored in the biomass into hot gaseous substances. Normally, in the furnace, steam turbine, or boiler at elevated temperature (above 800 °C), such types of process are adopted. Although it is possible for combustion of all kinds of biomass, the feasible and satisfactory combustion is only achieved in biomass products containing a lower percentage of moisture content. The combined effect of power and heat is highly necessary to enhance the efficiency of the overall plant.

The effectiveness of the net conversion of energy using biomass combustion for the power plants lies in the range of 20 to 40%. The efficiency can be enhanced in systems having power more than 100 MW by the co-combustion of biomass with coal. Therefore, it is less possible for the utilization of algal biomass in the method of direct combustion. However, the co-firing of coal-algae in their life cycle is the cause of less emissions of greenhouse gases. The method of direct combustion is generally carried out by the oxygenation of microalgae biomass in the steam turbine, boiler, or furnace and produces very hot gaseous components. Before direct combustion, the biomass products of the microalgae are first dried and then made into small particles of required size by grinding. In the technique of hydrothermal liquefaction, the algal slurries, if subjected to the pressure of 40–200 bar and temperature of 300–400 °C, produce crude biogas and coal char. Hence, different substances can be depolymerized and extracted from the biomass of microalgae through the process of liquefaction. The yield of oil obtained from hydrothermal liquefaction is normally in the range of 9-97 %, which is more than bio-oil production by pyrolysis [12, 13].

6.2.2 Biochemical Conversion

The conversion of microalgae biomass through biological methods into biofuel includes photobiological production of H_2, alcoholic fermentation, and anaerobic digestion. The biochemical path of transformation includes hydrolysis of the cell walls through bacteria into sugars of a fermentable state. The process of fermentation means the anaerobic digestion of biomass products into biogas, biohydrogen, and bioethanol. Biogas is generated through this method of acetogenesis in which the whole fermentable substances are oxidized into acetate and then converted into CH_4 and CO_2 by the process of methanogenesis. The production of biogas is affected by a C:N ratio of the biomass feedstock, time duration, rate of feeding, pH, temperature, and the amount of solids. The percent yield of biogas is less because of sensitivity to bacterial degradation and low C:N ratio, which produces NH_3 (serves as an inhibitor). The biomass product of Scenedesmus spp. without amino acid and lipids offers a better yield of biogas relative to the raw material. To enhance the C:N ratio, the biomass of algae is normally co-digested with sewage sludge and waste papers, which increases the production of CH_4 by 26%. The pre-treatment of algae biomass by using microwave irradiation enhances the yield of biogas by 56% due to the change in the structure of the cell walls. The rate of production of CH_4 can be increased through enzymatic, thermal, and mechanical

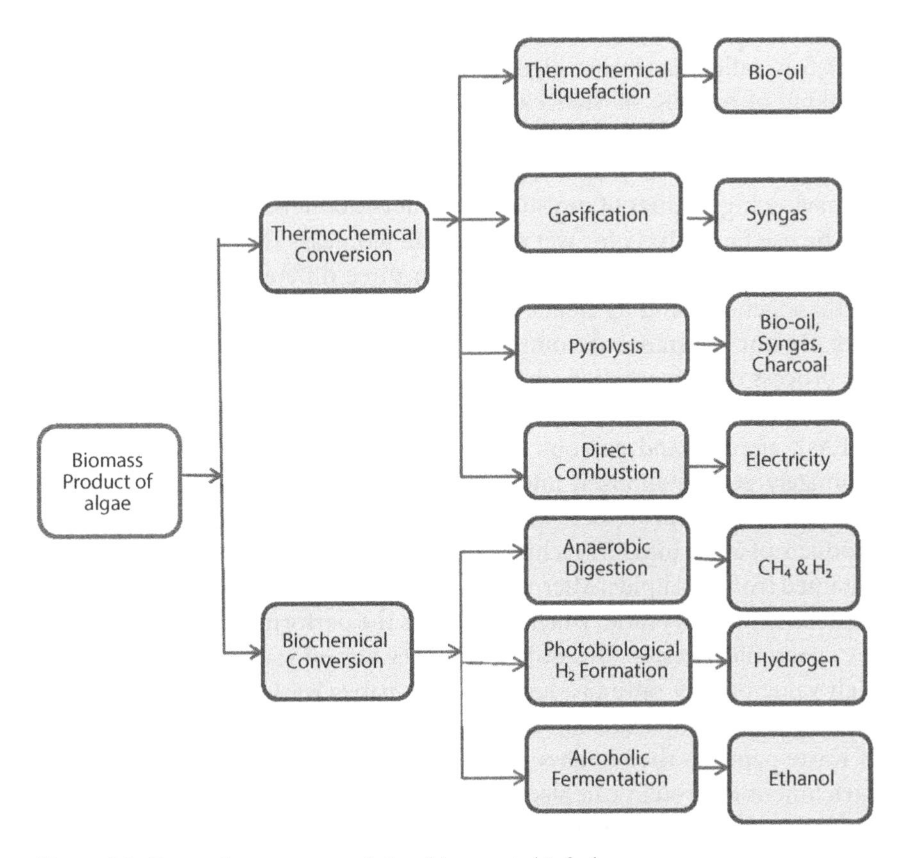

Figure 6.1 Conversion processes of algae biomass to biofuel.

pre-treatments. Bioethanol is derived due to fermentation of yeast from the hydrolysed carbohydrates. Phaeophyceae is treated as the most viable feedstock for the production of bioethanol because of the presence of a higher percentage of sugar in it. Various pre-treatments processes such as milling, liquefaction, hot water wash, saccharification, enzymatic hydrolysis, and alginate extraction are necessary for effective production of bioethanol. Photosynthesis by-products, such as O_2, can suppress the pathway of hydrogenase, whereas anaerobic digestion can overcome this issue. It was reported that there is a potential increase in the production of biohydrogen achieved (20 times) in a continuous flow regime over batch production [14, 15]. The potential conversion methods of algal biomass are shown in Figure 6.1.

6.2.2.1 Anaerobic Digestion

The process of conversion of organic waste materials into biogas which contain CH_4, CO_2 along with a small percentage of some other gases such as H_2S is called anaerobic digestion. This process generally involves the breaking of organic substances with the production of some simple gaseous substances. For this process, about 20–40 % less energy is required for heating. The method of anaerobic digestion is suitable for organic materials having high values of moisture content of about 80–90% and therefore can be used effectively for wet microalgae biomass. The method of anaerobic digestion is generally carried out in three different steps: hydrolysis, methanogenesis, and fermentation. In the method of hydrolysis, the complex organic substance decomposes into simple water soluble sugars. In the process of fermentation, the bacterial community converts the simple sugar molecules into other products such as CH_3COOH, volatile fatty acids (VFAs), alcohol, and gaseous substances containing CO_2 and H_2, which ultimately get metabolized into 60–70% of CH_4 and 30–40% of CO_2 via the methanogens process. It was identified that the conversion of biomass products of algae into CH_4 achieves higher energy compared to the energy obtained from cell lipids. Microalgae contains a high percentage of proteins because of less C/N ratio, which influences the performance and activity of the anaerobic digester. This difficulty can be solved via co-digestion with a high value of C/N ratio products, such as paper wastes, which is attained by a significant intensification in the production of CH_4 with the mixing of waste paper to the biomass product of algal. The algae biomass having enrichment of protein can also result in the enhancement of the production of ammonium, which decreases anaerobic microorganisms. Sodium ion is highly toxic to many anaerobic microorganisms, however it is viable to use the salt-adapted microorganisms for the anaerobic digestion of the biomass products of marine algae [16, 17].

6.2.2.2 *Alcoholic Fermentation*

The process of alcoholic fermentation is nothing but the conversion of biomass substances (containing sugars, cellulose, and starch) into alcohol (C_2H_5OH). The biomass product is first crushed and the starch obtained from it gets converted into sugars, which are then added to yeast and water. At a particular temperature (warm condition), fermentation occurs in the fermenter. The sugar present in algae biomass is broken down by yeast and converted into lower alcohol such as C_2H_5OH. The impurities and water can be removed from the dilute form of ethanol by the process of purification (distillation). More than 95% concentrated solution of ethanol vapours are condensed in the liquid state, which can be used as an alternative fuel or blended with petrol for use in any combustion engine. The solid by-products obtained from this method can be utilized for gasification purposes, for energy generation, and as cattle-feed. The biomass product of microalgae enriched with starch needs some additional unique processing technology before the fermentation process. The microalgae (C. vulgaris) are treated as an excellent source for the synthesis of C_2H_5OH and the efficiency of conversion of ethanol was found to be up to 65% because of the presence of high concentration of starch (dry basis: 37 %). Ethanol can also be produced from the biomass of microalgae through the process of dark fermentation and attained a maximum productivity of C_2H_5OH of 450 mmol/gm of dry weight at a temperature of 30°C. Hence, it can be concluded that the production of ethanol from the biomass product of microalgae is technically feasible, but cannot be commercialized because of many deficiencies [18, 19].

6.2.2.3 *Photobiological Hydrogen Production*

Hydrogen is considered as an efficient and clean carrier of energy that is found naturally. Microalgae are significant biological species, which possess highly required genetic, enzymatic, and metabolic characteristics in order to produce H_2. It was found that H_2 can be formed at anaerobic conditions from eukaryotic species of microalgae either as an electron donor in the process of CO_2 fixation or may grow in both dark and light phase. In the process of photosynthesis, the microalgae species convert H_2O molecules into H^+ ion and O_2. The H^+ ions are converted into H_2 by the enzyme hydrogenase in an anaerobic condition. Because of the reversibility of the chemical reaction, H_2 is either consumed or synthesized via the conversion of H^+ ion to H_2. The O_2 produced by the photosynthetic activity is the cause of fast inhibition to the vital hydrogenase and enzymes and the method of production of H_2 by photosynthetic activity is obstructed. Hence, the culture of microalgae for the purpose of the production of H_2 must be subjected to an anaerobic environment. There are usually two basic methods for production of photosynthetic H_2 from water and wastewater, where the photosynthetic production of H_2 and O_2 are separated spatially. In the first phase of this process, the growth of algae was carried out photosynthetically in an appropriate condition and in the second phase, the algae are deprived of sulphur by way of promoting an anaerobic environment and stimulating the constant production of H_2. Since after 60h of the production the yield of H_2 will start to stabilize, the process of production becomes restricted with the passage of time. The application of this system of production does not create any environmentally harmful or toxic substances, but rather forms valuable product materials because of biomass cultivation. The second method involves in the generation of photosynthetic H_2 and O_2. In this method, the electrons released due to photosynthetic oxidation of H_2O are directly subjected to the process of evolution of H_2 mediated hydrogenase. The production of H_2 seems to be advantageous theoretically in a two-stage photosynthetic approach, but the simultaneous creation process may suffer a great inhibition of hydrogenase even after a very small time period because of production of oxygen photosynthetically [20, 21].

6.3 Production of Biodiesel from Algal Biomass

Biodiesel is normally a derivative of some oil crops and biomass products and it can be utilized directly in usual diesel engines. It is a combination of various mono-alkyl esters having long chain fatty acids (FAME), which are extracted from algal oil. After the extraction of fatty acids from algal oil, it can be formed into biodiesel by the process of transesterification.

The process of transesterification is nothing but the reaction between an alcohol and triglycerides along with the addition of a suitable catalyst in order to produce mono-esters, which are ultimately called biodiesel. Since the properties of biodiesel produced from algae tallies with the international standard of biodiesel for vehicles, it can be accepted as a proper alternative of fossil fuels. Algal oils normally contain a higher percentage of polyunsaturated fatty acids compared to vegetable oils, which is the cause of resistance to oxidation during storage, therefore its use is restricted. The biodiesel obtained from algal biomass possesses similar chemical and physical properties as fossil petroleum diesel. Again, the biodiesel obtained from the algae biomass also exhibits several advantages compared to fossil fuel and petroleum diesel such as biodegradability, being derived from renewable sources, and being quasi-carbon neutral in the sustainable way of production. It is non-toxic, non-hazardous, and contains less particulate matter, CO, hydrocarbons, and SO_x, and produces less soot particles during combustion. The biodiesel derived from algal biomass is comparatively more efficient than 1st generation biodiesel for application in the aviation industry in which high energy densities and low freezing points are important criteria. Another significant benefit of algal biodiesel is the decrease in emissions of CO_2, almost up to 78% compared to the petroleum diesel [22, 23].

6.3.1 Transesterification

In this process, C_2H_5OH or CH_3OH reacts with triglycerides of fatty acids in the presence of either basic or acidic catalysts to produce glycerol and biodiesel. The rate of chemical reaction mainly depends upon the kinds of alcohol used, molar ratio, and nature of the catalysts added. This method is highly significant because it retards the viscosity and enhances the fluidity of algal oil to be mixed with fossil petroleum diesel for direct use in internal combustion engines. The process of traditional transesterification normally involves the extraction of lipids from algal biomass products and after that, the esterification process is followed. The process of direct transesterification is carried out in a single-step where the wet biomass product is used directly without any extraction procedures. The direct transesterification requires less energy, reagent, and time, but percent of yield is less as compared to the conventional treatment process. In order to save energy, the most suitable approach is the combined effect of ultrasound and microwave irradiation techniques. As a result, the yield can be enhanced up to 90%. The process of the in-situ supercritical CH_3OH transesterification process is another important path that can be used for a reduction in the cost of production. The different fatty acids which are mainly used for producing improved quality of biodiesel include palmitic acid (C16:0), oleic acid (C18:1), stearic acid (C18:0), linolenic acid (C18:3), and linoleic acid (C18:2). The properties of biodiesel are highly dependent upon the content of fatty acids present in it. Polyunsaturated fatty acids (PUFAs) are usually hostile due to their susceptibility towards oxidation and the occurrence of lipids in its saturated state. The saturated lipids have a tendency to increase the cloud point, viscosity, and pour point of the derived biodiesel. Hence, the monounsaturated fatty acids (MUFAs) are always the most preferred lipids. The heat of combustion, viscosity, and melting point of the biodiesel are solely enhanced to increase the chain length of the fatty acid and are inversely proportional to the unsaturation. The lubrication property and autoxidation enhances with an increase in unsaturation of the fatty acids. Lipase acts as a catalyst in the method of transesterification and, simultaneously, catalyse processes and promotes the recovery of the by-products. The algae chlorophytes and cyanobacteria both possesses a higher content of lipids and therefore, exhibits high capability in the production of biodiesel. The algae Oleaginous strains contain a minimum 20% of lipid in dry basis which can produce lipids up to about 70% on the dry weight basis in particular conditions of massive stress such as Si or/and N_2 deprivation [24–26].

6.4 Genetic Engineering Toward Biofuels Production

The techniques of genetic engineering are particularly stable in genetic transformation and heterologous gene expression and needs for the production of developed algal biofuel of a high quality. The gene Acetyl-CoA carboxylase (ACCase) is usually suitable for catalysing during the first step synthesis of fatty acids. However, the overexpression of this kind of gene alone is not sufficient to increase the production of oil in algae. The gene ACCase are normally overexpressed in Cyclotella cryptica, which is the cause of a small increase in the percentage of lipid.

A potential enhancement in the algal lipids is normally observed due to the overexpression of the compound diacylglycerol acyltransferase (DAGAT). In addition to that, C. reinhardtii is found to accumulate a higher concentration of lipid. If the metabolic path of biopolymer starch was obstructed, a comparable result was found for Chlorella pyrenoidosa. The concentration of lipid would be influenced by stimulating ACCase and DAGAT with obstruction of the pathway of starch and enhances the synthesis of the lipid.

The increase in the production of H_2 through the pathway of genetic engineering is the cause of the retardation of light antenna harvesting size and also decreases with the hydrogenase concentration. The distorted strain of the gene Chlamydomonas reinhardtii (Stm6), along with a blocked cyclic flow of electron via Photosystem, exhibits the enhanced accumulation of starch along with the depletion of intracellular concentrations of O_2 or inhabitation of hydrogenase.

The process of anaerobiosis can be accomplished through a Cu receptive nuclear transgene. Hence, in order to overcome the demerits of penetration, a single RNAi construct was capable of successfully silencing the whole light-harvesting complex (LHC) protein, which is the isoforms of C. reinhardtii, and causes enhancing transmittance of light in the culture by almost 290%.

The growth and culture of algae under the mixotrophic or heterotrophic condition is the cause of higher cell densities, which results in a decrease in the cost of harvesting. It was observed that most of the algae are autotrophic in nature. The various algae, including Volvox carteri, Cylindrotheca fusiformis, C. reinhardtii, and Phaeodactylum tricornutum, were efficiently transformed through a hexose transporter (HUP1) resulting in transport of glucose into the cells. Although genetic engineering is considered as advanced technology for biodiesel production, it is still not economically viable and needs some modification [27–29].

6.5 Summary

From this study it can be summarised that as per the requirement and the types of algae available, the process of conversion can be chosen for the production of biofuel. All the described processes are suitable for the conversion of algae to biofuels and chemicals. However, more research is required to find the best type of bio-reactors and more efficient processing conditions to produce more algae biofuels. Among all the processes, the Phototrophic production process is more viable and effective than heterotrophic production. The process of transesterification and genetic technology in the production of biofuel from algae is now considered the most significant technology. The mitigation of the emissions of CO_2 with microalgae offers a balancing function, which is another cause of economic viability of the production of biofuel. The cultivation of algal biomass is an important step in biofuel production and until now, no standard effective technology is being adopted for this purpose. Lipids are considered the most suitable feedstock of biofuel from algae, but the major limitation is the existence of polyunsaturated fatty acids (PUFAs), which undergo oxidation and a higher percentage of moisture in the feedstock of algae. Finally, we can say that both pyrolysis and thermochemical liquefaction are the most technically viable approaches for the conversion of algal biomass to biofuel after the removal of oils from algae.

References

1. Halder, P., & Azad, A. K., Recent trends and challenges of algal biofuel conversion technologies. *Advanced Biofuels,* Chapter-7, 167–179(2019). doi:10.1016/b978-0-08-102791-2.00007-6
2. Chen, W.-H., Lin, B.-J., Huang, M.-Y., & Chang, J.-S., Thermochemical conversion of microalgal biomass into biofuels: A review. *Bioresource Technology,* 184, 314–327(2015). doi:10.1016/j.biortech.2014.11.050
3. Chiaramonti, D., Prussi, M., Buffi, M., Casini, D., & Rizzo, A. M., Thermochemical Conversion of Microalgae: Challenges and Opportunities. *Energy Procedia,* 75, 819–826(2015). doi:10.1016/j.egypro.2015.07.142
4. Gollakota, A. R. K., Kishore, N., & Gu, S. A review on hydrothermal liquefaction of biomass. *Renewable and Sustainable Energy Reviews,* 81, 1378–1392(2018). doi:10.1016/j.rser.2017.05.178
5. Raheem, A., Wan Azlina, W. A. K. G., Taufiq Yap, Y. H., Danquah, M. K., & Harun, R., Thermochemical conversion of microalgal biomass for biofuel production. *Renewable and Sustainable Energy Reviews,* 49, 990–999(2015). doi:10.1016/j.rser.2015.04.186
6. Azizi, K., Keshavarz Moraveji, M., & Abedini Najafabadi, H., A review on bio-fuel production from microalgal biomass by using pyrolysis method. *Renewable and Sustainable Energy Rev*iews, 82, 3046–3059(2018). doi:10.1016/j.rser.2017.10.033
7. Lee, S.Y., Sankaran, R., Chew, K.W. *et al.* Waste to bioenergy: a review on the recent conversion technologies. *BMC Energy* 1(1), 4-25 (2019).
8. Ethaib, S., Omar, R., Kamal, S. M. M., Awang Biak, D. R., & Zubaidi, S. LMicrowave-Assisted Pyrolysis of Biomass Waste: A Mini Review. *Processes,* 8(9), 1190-1216 (2020). doi:10.3390/pr8091190
9. Zewdie, D.T., Ali, A.Y. Cultivation of microalgae for biofuel production: coupling with sugarcane-processing factories. *Energy, Sustainability and Society,*10, 27-42 (2020). https://doi.org/10.1186/s13705-020-00262-5
10. Yang, C., Li, R., Cui, C., Liu, S., Qiu, Q., Ding, Y., Zhang, B. Catalytic hydroprocessing of microalgae-derived biofuels: a review. *Green Chemistry,* 18(13), 3684–3699(2016).doi:10.1039/c6gc01239f
11. Faruque, M. O., Razzak, S. A., & Hossain, M. M., Application of Heterogeneous Catalysts for Biodiesel Production from Microalgal Oil—A Review. *Catalysts,* 10(9), 1025-1049(2020). doi:10.3390/catal10091025
12. Kumar, G., Shobana, S., Chen, W. H., Bach, Q. V., Kim, S. H., Atabani, A. E., & Chang, J. S., A review of thermochemical conversion of microalgal biomass for biofuels: Chemistry and processes. *Green Chemistry,* 19(1),44-67(2017). https://doi.org/10.1039/c6gc01937d
13. Milledge, J. J., & Heaven, S., Methods of energy extraction from microalgal biomass: a review. *Reviews in Environmental Science and Bio/Tec*hnology, 13(3), 301–320(2014). doi:10.1007/s11157-014-9339-1

14. Hossain, S. M. Z., A Tutorial Review on Biochemical Conversion of Microalgae Biomass into Biofuel. *Chemical Engineering & Technology*. 42(2), 2594-2607(2019).doi:10.1002/ceat.201800605
15. Choudhary, P., Assemany, P. P., Naaz, F., Bhattacharya, A., de Siqueira Castro, J., de Aguiar do Couto, E., Malik, A., A review of biochemical and thermochemical energy conversion routes of wastewater grown algal biomass. *Science of The Total Environment*, 726, 1-24(2020), 137961. doi:10.1016/j.scitotenv.2020.137961
16. Wirth, R., Lakatos, G., Böjti, T., Maróti, G., Bagi, Z., Rákhely, G., & Kovács, K. L. Anaerobic gaseous biofuel production using microalgal biomass – A review. *Anaerobe*, 52, 1–8(2018). doi:10.1016/j.anaerobe.2018.05.008
17. Milledge, J., Nielsen, B., Maneein, S., & Harvey, P. A Brief Review of Anaerobic Digestion of Algae for Bioenergy. *Energies*, 12(6), 1166-1187(2019). doi:10.3390/en12061166
18. Dave, N., Selvaraj, R., Varadavenkatesan, T., & Vinayagam, R., A critical review on production of bioethanol from macroalgal biomass. *Algal Research*, 42, 1-14(2019).101606. doi:10.1016/j.algal.2019.101606
19. De Farias Silva, C. E., & Bertucco, A., Bioethanol from microalgae and cyanobacteria: A review and technological outlook. *Process Biochemistry*, 51(11), 1833–1842(2016).doi:10.1016/j.procbio.2016.02.016
20. Khetkorn, W., Rastogi, R. P., Incharoensakdi, A., Lindblad, P., Madamwar, D., Pandey, A., & Larroche, C. Microalgal hydrogen production – A review. *Bioresource Technology*, 243, 1194–1206(2017). doi:10.1016/j.biortech.2017.07.085
21. Khosravitabar, F. Microalgal biohydrogen photoproduction: scaling up challenges and the ways forward. *Journal of Applied Phycology*, 32, 277–289 (2020). https://doi.org/10.1007/s10811-019-01911-9
22. Patle, D.S., Pandey, A., Srivastava, S. *et al.* Ultrasound-intensified biodiesel production from algal biomass: a review. *Environ Chem Lett* (2020). https://doi.org/10.1007/s10311-020- 01080-z
23. Karmakar, R., Rajor, A., Kundu, K. *et al.*, Production of biodiesel from unused algal biomass in Punjab, India. *Petroleum Science*, 15, 164–175 (2018). https://doi.org/10.1007/s12182-017-0203-0
24. Akubude, V. C., Nwaigwe, K. N., & Dintwa, E., Production of biodiesel from microalgae via nanocatalyzed transesterification process: A review. *Materials Science for Energy Technologies*. 2, 216-225(2019)
25. Makareviciene, V., & Skorupskaite, V. Transesterification of microalgae for biodiesel production. *Second and Third Generation of Feedstocks, Chapter—17,* 469–510. (2019) doi:10.1016/b978-0-12-815162-4.00017-3
26. Hidalgo, P., Toro, C., Ciudad, G. *et al.*, Advances in direct transesterification of microalgal biomass for biodiesel production. *Rev Environ Sci Biotechnol* 12, 179–199 (2013). https://doi.org/10.1007/s11157-013
27. Tabatabaei, M., Tohidfar, M., Jouzani, G. S., Safarnejad, M., & Pazouki, M. Biodiesel production from genetically engineered microalgae: Future of

bioenergy in Iran. *Renewable and Sustainable Energy Reviews*, 15(4), 1918–1927 (2011). doi:10.1016/j.rser.2010.12.004

28. Radakovits, R., Jinkerson, R. E., Darzins, A., & Posewitz, M. C. Genetic Engineering of Algae for Enhanced Biofuel Production. Eukaryotic Cell, 9(4), 486–501(2010). doi:10.1128/ec.00364-09
29. Lin, H., Wang, Q., Shen, Q., Zhan, J., & Zhao, Y. Genetic engineering of microorganisms for biodiesel production. *Bioengineered*, 4(5), 292–304(2013). doi:10.4161/bioe.23114

7

Technologies of Microalgae Biomass Cultivation for Bio-Fuel Production: Challenges and Benefits

Trinath Biswal[1], Krushna Prasad Shadangi[2*] and Prakash Kumar Sarangi[3]

[1]Department of Chemistry, Veer Surendra Sai University of Technology, Burla, Odisha, India
[2]Department of Chemical Engineering, Veer Surendra Sai University of Technology, Burla, Odisha, India
[3]College of Agriculture, Central Agricultural University, Imphal, Manipur, India

Abstract
The utilization of fossil fuel contributes a significant fraction of greenhouse gas to the troposphere. Now, rigorous research initiatives are carried out to find out the alternative sources to fossil fuels, which will be sustainable and can be commercialized to meet the market demand. In this situation, the biofuel obtained from renewable sources is the only alternative in global economy from a sustainable point of view. The biofuel cannot be produced from crop grains as it may create a scarcity of food worldwide. The biofuel derived from algae is a proper alternative to fossil fuels, however the technology adopted must be modified in order to overcome the demerits associated with it to compete with the fuel market. This chapter elaborates on the details of the algae cultivation methodology, benefits of the use of algae as a source of biofuel production, challenges associated with it, and the impact of microalgae on the environment.

Keywords: Microalgae, biomass, greenhouse gas, biofuel, sustainable source

**Corresponding author*: krushnanit@gmail.com; kpshadangi_chemical@vssut.ac.in

Prakash Kumar Sarangi (ed.) Biorefinery Production of Fuels and Platform Chemicals, (147–178) © 2023 Scrivener Publishing LLC

7.1 Introduction

Sustainability is considered the crucial principle for the management of natural resources, which is associated with minimization of ecological impact, operational efficiency, and socio-economic conditions and all are interdependent from each other. Energy resources based on fossil fuels continuously deplete at a rapid rate and this is the most important issue among these liquid fossil fuels. Almost 88% of the total energy production is from fossil fuels, among which the share of petroleum oil is 35%, coal 29%, natural gas 24%, hydroelectricity 6%, and nuclear energy 5% [1, 2]. The direct use of fossil fuels is the cause of many negative impacts on the environment, such as air, water, and soil contamination. Fossil fuels are the major contributor to greenhouse gases and a potential threat to the biological community of the world. Hence, biofuel is considered a promising alternative to fossil fuels [3]. The first-generation biofuels are extracted from various terrestrial crops, including sugar cane, maizes, sugar, and beets and their use in biofuel production is the cause of enormous strain on the food market of the globe, as well as destruction of the forest areas of the globe. The second-generation biofuel is extracted from lignocellulosic biomass of forest residues, feedstock of non-food crops, and agricultural crops, which is the problem in proper land use and management [4, 5]. Hence, to overcome all these demerits, the third-generation biofuels extracted from microalgae are treated as the feasible alternative to fossil fuel. Microalgae are usually photosynthetic microorganisms and require light, N, sugars, K, P, and CO2 for their growth and reproduction. Microalgae synthesized lipids, carbohydrates, and proteins in huge quantities in a short interval of time and these products formed can produce biofuel and some other necessary by-products. Careful use of microalgae can meet the primary demand of energy while simultaneously benefitting crops from an environmental point of view [6, 7]. The growth potential of microalgae is very high and many species of it have an oil content of about 20–50% (dry weight of biomass) and the rate of exponential growth becomes two times their biomass in the time period of 3 1/2 h. Nutrients such as P and N for the cultivation of microalgae can be found from waste water and sewage water, so besides the growth medium there is a dual potential observed for organic effluent treatment obtained from the agro-food industry. The cultivation of algae does not need any pesticides or herbicides [8, 9]. The algae biomass produces valuable products such as lipids and proteins and the residual product obtained after extraction of the oil can be effectively used as fertilizer or fermented to synthesize CH4 and

C2H5OH. The microalgae is also able to synthesize biohydrogen by the photobiological method. Besides biofuel production and CO2 fixation, the production of biohydrogen and biotreatment of wastewater effluents are the other significant applications of microalgae [10, 11].

7.2 Challenges Towards Algae Biofuel Technology

In spite of the many potential benefits of biofuel resources, there are still many challenges that have been obstructing development. The modification of this technology for the production of algal biofuels to make it commercially feasible for government permits and path of utilization in a sustainable manner is required. Some of the challenges are as follows:

- The proper selection of algae species, which balances the necessity of production of biofuel and formation of valuable by-products
- For achieving efficiencies of higher photosynthetic activity, the continuous modification of production systems is required
- Development and modification of the techniques for the cultivation of single species and diffusion losses due to CO_2
- Negative energy balance potential subsequently accounting for the necessities in water pumping, harvesting, transfer of CO_2, and extraction
- There are few numbers of commercially viable plants that are now in operation, hence there is a shortage of data for large scale plants
- The introduction of flue gases, which are inappropriate in high concentration because of the existence of dangerous substances such as SO_x and NO_x

Sustainability is considered as the vital natural resource of exploitation or management and normally involves environmental, operational, and socio-economic issues, where all are mutually dependent upon each other. The higher rate of growth, high percentage of oil, and practicable growth densities are the cause of inducing factors for the production of biofuels from algae biomass products. However, the growth of algae made it suited for the production of biofuel in an economical way and also creates many hurdles. Hence, finally, it can be suggested that the major challenges of biofuel production from algal biomass include isolation of strain, source of nutrients, and its utilization [12–14].

7.3 Biology Related with Algae

Algae are identified as one of the oldest life-forms on our globe. Microalgae normally belong to the primitive class of plants or called thallophytes. These plants are not strong in roots, leaves, and stem without any sterile covering of the cells around or neighbouring to the reproductive cells and are considered as a basic photosynthetic pigment. The structure of algae is basically used for the production of energy without any kind of development [15].

Prokaryotic species or cyanobacteria are deficient in membrane-bound organelles such as plastics, nuclei, Golgi bodies, flagella, and mitochondria. These are found to be more analogous to bacterial species rather than algae. The eukaryotic species contain various kinds of common algae. All these organelles can be controlled and regulated by the functions of the cell, facilitating survival and reproduction. The eukaryotic species are generally divided into various kinds of classes that depend on different factors such as the nature of pigmentation, basic cellular structure, and life cycle [16].

The most significant classes of algae used for the biofuel synthesis are as follows:

- Chlorophyta, commonly called green algae
- Rhodophyta, commonly called red algae
- Bacillariophyta or diatoms algae.

Algae may be either being heterotrophic or autotrophic organisms. The autotrophic organism needs only inorganic substances. These are salts, CO_2, and light energy that are necessary for their existence and growth. The autotrophic organisms are non-photosynthetic in nature and require an outside source of organic materials such as a carbon source (food) and nutrient availability (as a source of energy).

Some of the photosynthetic algae are found to be mixotrophic in nature and therefore have the capability to achieve exogenous organic nutrients and photosynthesis. In the case of an autotrophic class of algae, photosynthesis is the vital factor for their existence and survival, where they convert the absorbed CO_2 and sunlight through chloroplasts into oxygen and adenosine triphosphate (ATP). Further, it is utilized for respiration by generating energy and promoting growth [17, 18].

Microalgae are found to be a prominent source of biofuel, which are naturally obtained from the marine ecosystem and freshwater bodies. Microalgae are treated as one of the oldest and most significant living microorganisms present in our globe and occupy the bottom part of the food chain. Normally, more than 300,000 kinds of microalgae are found in our globe, whose diversity is superior compared to the various plant species. This belongs to the halophytes categories, which means plants deficient in root systems, stems, and leaves (contains chlorophyll as a basic photosynthetic pigment) and deficiency of sterile covering of the cells around the reproductive cells [19].

The mechanism of photosynthetic activity of microalgae is almost similar to that of trees and plants, therefore microalgae can convert solar energy to their simple cellular form of the structure. The cells grown are comfortable in the suspension of water because these are comparatively more efficient for accessing H_2O, $CO_{2,}$ and nutrients. Presently, microalgae are classified according to their colour, types of pigments, chemical properties of the storing products, and constituents present in the cell wall [20].

Some extra criteria are also considered for the classification of microalgae. These are classified per their morphological characteristics, cytological properties, flagellate cells pathway, cell structure, scheme of the nuclear division, cell division, and the existence of an envelope size of endoplasmic reticulum around the chloroplast with probable interlinking in between nuclear membrane and endoplasmic reticulum [21]. Some major microalgae on the basis of their colour are shown in Table 7.1 and the various

Table 7.1 Groups of microalgae on basis of colour.

Colour	Groups
Yellow-green algae	Xanthophyceae
Cyanobacteria	Cyanophyceae
Red algae	Rhodophyceae
Brown algae	Phaeophyceae
Golden algae	Chrysophyceae
Green algae	Chlorophyceae

Table 7.2 Contents of oil present in microalgae.

Name of microalgae	Dry wt. (%)
Botryococcus brauni	25 75
Tetraselmis sueica	15 - 23
Chlorella sp	28 32
Schizochytrium sp.	50-77
Crypthecodinium cohnii	20
Phaeodactylum tricornutum	20-30
Cylindrotheca sp	16-37
Nitzschia sp.	45-47
Dunaliella primolecta	23
Neochloris oleoabundans	35-54
Isochrysis sp	25-33
Nannochloropsis sp.	31-68
Monallanthus salina	20
Nannochloris sp.	20-35

species of the microalgae in association with their dry weight are represented in Table 7.2 [22].

Hence, it was confirmed that various species of microalgae possesses oil concentration up to around 80% of the dry weight of their body. There are some species of microalgae whose mass becomes doubled within a day. The shortest time of doubling of biomass is of about 3-4 hours, which establishes the microalgae as a model renewable source for the production of biofuel. Microalgae are generally unicellular microorganisms with photosynthetic activity that colonize in marine and freshwater ecosystems. All kinds of algae significantly accumulate molecules rich in energy content, including polysaccharides and oil. In addition to that, microalgae have the capability to produce biomass rich in protein, which depends upon environmental conditions and kinds of algae species. Microalgae based biofuel is more attractive since it can be cultivated even in non-arable land. Microalgae is the cause of offering a number of high-value materials including pigments, antioxidants, long-chain polyunsaturated acids (mainly mega-6 and omega-3), volatiles, extracts for cosmetics, and many substances having commercial importance. The metabolically engineered microalgae can be successfully utilized to produce a high value of product, such as terpenoids, from CO_2 or some other heterologous materials [23, 24].

7.4 Algae Biofuels

The fuel extracted from algae plant biomass is called algae biofuel. Since the feedstock can be refilled easily, biofuel is considered as a potential source of renewable energy [25]. The whole life cycle of algae is represented in Figure 7.1. The biofuel is now treated as an eco-friendly, economical, and sustainable alternative to fossil fuels (coal and petroleum products) and especially a viable solution to reduce the demand and price of petroleum products and to reduce global warming. Biodiesel is the second most widely available popular liquid biofuel, which is mainly extracted from plant species enriched in oil such as palm or soybean oil. Biofuel was also obtained from less oily waste products such as cooking fat wastes from hotels and

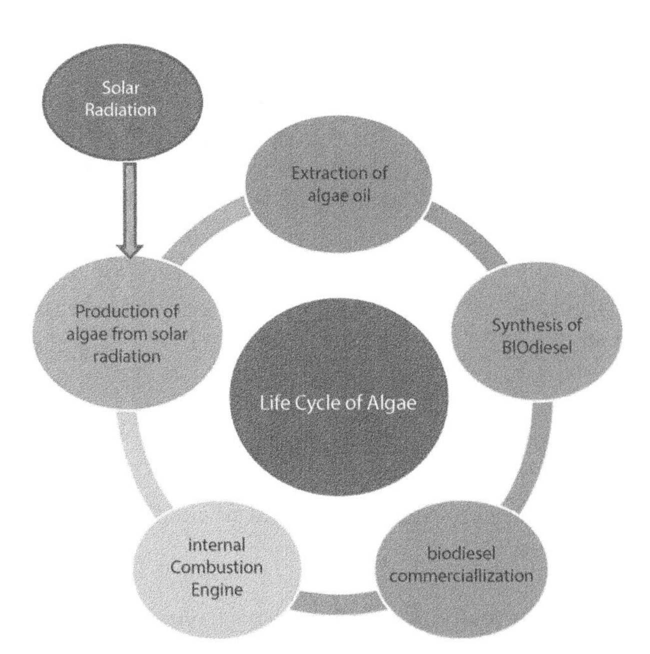

Figure 7.1 Life cycle of algae.

restaurants after deep-frying. Biofuel is mainly used in European countries for blending with diesel or other petroleum oil in appropriate percentages. The biofuel extracted from cyanobacteria and algae are considered as a potential source of third-generation biodiesel, which holds a promising approach, but its development and synthesis is difficult and costly [26].

Some species of algae contain about 40 wt. % of lipids and can be synthesized into synthetic petroleum or biodiesel. It was assessed that cyanobacteria and algae produces about 10 to 100 times more biofuel per unit area as compared to the second-generation biofuels. Crops like sugarcane and corn globally serve as the potential sources for bioethanol production. Bioethanol is a first generation biofuel and is unable to transport on CO_2 mitigation targets. It also contributed to a host of socioeconomic and environmental issues of the world. This is because of the vast population, availability of limited area of arable land, and very high demand of energy. First generation biofuels normally have less power density and require more arable land, which is practically not possible for European countries. The increase in the cultivation of crops for the purpose of biofuel partially can fulfil the demand of fuel in the developed countries along with the food requirements of the developing countries of the world [27].

This fuel versus food dilemma exaggerates both these and is partly accountable for the increase in the food crisis of the world, becoming the cause of extensive starvation and malnutrition along with social and political conflict. Recently, it was identified that many first generation biofuels are the cause of CO_2 rising in the atmosphere if the fertilizer, land, and energy use are calculated for its production. There are some cases where the emission of CO_2 is less, but not economical. Simple microorganisms having photosynthetic activity, mainly cyanobacteria and green algae, are now becoming a significantly viable source of biofuel. Since all the crude oils are finally obtained from algae, it positively becomes the major source of fuel in future [28, 29].

7.5 Benefits of Microalgal Biofuels

Biofuel derived from micro-algae has several advantages and some of them are summarised here.

- Algae are normally aquatic organisms having various species ranging in size from microalgae to macroalgae. Different options are selected to recognize the kinds of algae and strain choice. These vary from a single cell to multicellular structures such as filaments and colonies.

- Microalgae are considered single-celled organisms of diversified classes which can potentially provide a viable solution regarding the scarcity and suitability of liquid fuel used in transportation in a number of ways.
- Different categories of algae species normally grow and reproduce comfortably in a wide range of aquatic ecosystems including fresh water, saturated saline water, and marine water.
- Algae use CO_2 and account for more than 40% of the total global carbon fixation. The majority of this productivity normally comes from marine microalgae.
- Algae can be synthesized in a very rapid process and also have the capability to produce energy-rich oils and a variety of kinds of microalgae that can naturally accumulate themselves in the dry mass of substances possessing a high concentration of oil.

  Example: Certain Botryococcus spp. which have been recognized that about 50% of long-chain hydrocarbons are stored in dried form.
- The variety of kinds of species of microalgae are now being identified as prospective biofuel crops invented from groups whose family associations are potentially broader than many of the diverse land plants, which offers prosperity of genetic diversity. These groups of algae are mainly diatoms, golden brown, prymnesiophytes, green algae, cyanobacteria, and eustigmatophytes. The membranes of all these species of microalgae are verified as a potential source of fuel production strains. Cyanobacteria are not athealgae, but actually a group of photosynthetic bacteria [30–32]. Table 7.3 shows the experimental conditions for the production of biofuel from algae.
- Microalgae have some extra benefits over plant species in the terrestrial ecosystem. Since these are normally single-celled organisms, they can be duplicated by division and by using some advanced modified technologies to get the quick of strains.
- The use of algal biofuel reduces the time required for the production of algae from years to months. Again, use of

Table 7.3 Algal species used for production of biofuel with experimental conditions [33].

Fuel kinds	Algal species	Experimental conditions
Biodiesel	Oscillatoria sp.	BG-11 medium with varying concentrations of NO_3^-; (1500, 375, 186, 94, 47, 23 and 0.0) mgL^{-1} $NaNO_3$
	Chlorella sp.	60 °C, H_2SO_4, 4 hour, CH_3OH medium
	Chlorella sp. and Desmodesmus quadricaudatus	70 °C, H_2SO_4, 3hour, CH_3OH medium
	Chlorella pyrenoidosa	90 °C, H_2SO_4, 2hour, CH_3OH medium
	Chlorella sp. and Desmodesmus quadricaudatus	Pure batch cultures, nitrogen-free medium, BG-11 standard and hexane-ether, CH_3OH medium
	Dunaliella tertiolecta	110 °C, H_2SO_4, 5hour, CH_3OH-THF
	Dictyochloropsis splendida	110 °C, NaOH, 5hour, CH_3OH medium
	Nannochloropsis oculata	80 °C, NaOH, 2hour, CH_3OH chloroform (10:1)
	Schizochytrium limacinum	90 °C, H_2SO_4, 40min, $CH_3OH/CHCl_3$
	Spirulina sp	Concentration of catalyst, CH_3OH (80 mL), reaction time = 650 rpm, 8hour at 65 °C
Ethanol	Chlorella vulgaris	Pellet washed with 95% methanol (incubated with α- amylase (100 °C and pH =6) and glucoamylase (60 °C and pH =4.5) fermented by Saccharomyces cerevisiae (IFO-309), pre-treatment (ultrasonic radiation)

(Continued)

Table 7.3 Algal species used for production of biofuel with experimental conditions [33]. (*Continued*)

Fuel kinds	Algal species	Experimental conditions
Biohydrogen	Chlamydomonas reinhardtii	Anaerobic and aerobic environments, mixing speed of 170 ± 10 rpm/2.5min with light intensity (70 × 2 mmol/m²/s)
	Anabaena cylindrical	Pre-treatment with amylase followed by thermophilic fermentation under light intensity of 120 mmol/m²/sec
	Mastigocladus laminosu	Sparging the cultures with a gaseous mixture of 0.6% CO_2, 0.2 to 0.4% N_2, and rest Ar gas flows at rate of 3 L/hour
Bioethanol	Chlorococcum sp.	Yeast powder, 200 rpm, 30 °C, 60 hour, without pre-treatment
	Chlamydomonas reinhardtii	Pre-treatment by enzyme, 30 min,70–100 °C, rotation of 160 rpm, S. cerevisiae S288C anaerobically cultured at 30°C for 40 hour
Methane(CH_4)	Euglena gracilis	150 mmol/m²/s at 30°C
	Chlorella vulgaris	308°C, Batch culture, 30 days, pre-treatment (heat, 40 min/alkali)
	Arthrospira maxima	308°C, Continuous Flow Stirred-Tank Reactor, 2–4 days, pre-treatment process with magnetic stirring, and finally dried
	Spirulina sp.	308°C, Batch reactor, 28 days, without pre-treatment
	Scenedesmus obliquus	306°C, Anaerobic membrane bioreactor, 20 days, without pre-treatment

(*Continued*)

Table 7.3 Algal species used for production of biofuel with experimental conditions [33]. (*Continued*)

Fuel kinds	Algal species	Experimental conditions
Oil/gas/char	Tetraselmis Chuii	IR-pyrolysis, 500°C, fixed bed reactor, 10 °C/min
	Chlorella vulgaris	Closed tubular photobioreactor fluidized bed reactor, Fast pyrolysis, 500°C
	Synechococcus	Pyrolysis, 10°C/min, 500°C
	Dunaliella tertiolecta	Pyrolysis, 10°C/min, fluidized bed, 500°C
	Nannochloropsis sp.	Pyrolysis, 10°C/min, fixed bed reactor without or with HZSM-5, 400°C
Oil	Microcystis aeruginosa	Fast pyrolysis, 500°C, 10°C/min
	Chlorella protothecoides	Slow pyrolysis, 550°C, tubular reactor
Gas	Emiliania huxleyi	Pyrolysis, fixed bed reactor, 400°C temperature, batch cultivation
Bio-oil	Desmodesmus sp.	5 min at temperature 375°C
	Chlorella sp.	90 min at temperature 300°C
	Cyanobacteria sp.	45 min at temperature 325°C
	Chlorella vulgaris	60 min at temperature 300°C
	Bacillariophyta sp.	60 min at temperature 325°C
	Chlorogloeopsis fritschii	60 min at temperature 300°C
	Tetraselmis sp.	5 min at temperature 300°C
	Nannochloropsis sp.	90 min at temperature 300°C
	Spirulina platensis	60 min, 300°C

(*Continued*)

Table 7.3 Algal species used for production of biofuel with experimental conditions [33]. (*Continued*)

Fuel kinds	Algal species	Experimental conditions
	Nannochloropsis oculata	60 min at temperature 350°C
	Nannochloropsis gaditana	5 min at temperature 300°C
Syngas	Tetraselmis sp.	Temperature of 850°C, Fixed bed reactor, co-gasification (90% coal and 10% algae)
	Chlorella vulgaris	450°C, using batch reactor, 30 min
	Saccharina latissimi	450°C, 30min, using batch reactor, NaOH, Ni catalyst
	Nannochloropsis sp.	Fixed bed reactor, 1e10 bar, temperature 700–1000°C, 10,000°C/1 min
	Spirulina platensis	Ru/C, >400°C, Ru/ZrO_2
	Nannochloropsis oculat	Fixed bed reactor, 15 min, 850°C, Fe_2O3, CO_2 emission
	Nannochloropsis gaditana	Using TGA, temperature of 850°C,

algae significantly decreases the negative impact on the environment over other biomass products.

- ❖ Algae can grow and reproduce even in a land where traditional agriculture is not possible to grow. Hence, the production of biofuel from algae not only minimizes the use of agricultural land, but also remediated the waste streams. The various significant waste streams are normally urban wastewater effluents such as PO_4^{3-} and NO_3^-.
- ❖ Algae cultivation removes such effluents as flue gas produced due to coal burning and other combustible gases produced from the power plants by trapping the CO_2 and SO_4^{2-}
- ❖ Algae formation strains have also been potentially used in bioengineering in order to permit development of some specific traits along with the production of many precious

by-products. This makes the algal biofuels become economic and able to compete with petroleum fuels in the global market.
- For the existence and growth, algae requires more light energy to grow at a rapid rate and produce 2–15 times more lipids than any of the higher oil enriched plants such as Jatropha, soybean, and rapeseed [34–36].

7.6 Technologies for Production of Microalgae Biomass

At normal conditions, the growth of phototrophic algae absorbs sunlight and assimilates CO_2 from atmospheric air and nutrients from the aquatic environments. Hence, the artificial production may enhance and replicate the optimal natural growth circumstances. If the natural environment is used for commercial production of algae, then the benefit is the free use of natural resources and solar energy. However, this process is restricted only by the available sunlight in the daytime and the variation of light intensity with the season. For the commercial production of microalgae biomass, high intensity solar radiation is required. For outdoor production of algae, normally light is the limiting factor. For farming of phototrophic algae on a large scale, exclusively fluorescent lamps are used to fulfil the scarcity of light [37, 38].
Example: Diatoms of algae normally contain photosynthetic pigments, which include chlorophylls c and a, along with fucoxanthin, whereas the green algae normally contains chlorophylls b and a, along with zeaxanthin. Microalgae have the capability of fixing CO_2 from various sources that include:

- Atmospheric CO_2
- CO_2 present in the gases from large, medium, and small scale industries
- CO_2 obtained from the soluble carbonate salts [39].

Under the natural environment, microalgae assimilate CO_2 from the atmospheric air and most of the groups of microalgae can utilize and tolerate considerably high levels of CO_2 up to 150,000 ppm. Hence, the common method of addition of the CO_2 for the growth of algae is either from the gases eliminated from power plants or from carbonate salts such as

$NaHCO_3$ and Na_2CO_3, which are soluble in water. There are some inorganic nutrients potentially needed for the production of algae, including N, Si, and P. However, some of the species of algae have the capability of fixing N_2 from the atmospheric air to NO_x. Most of the microalgae need a soluble form of N_2, especially in the form of urea. Phosphorus is one of the important nutrients, which requires very smaller quantities for suitable growth and production of algae. However, it must have be supplied in a sufficient amount since PO_4^{3-} made a bond with the metal ions, therefore not all phosphorus supplied is bioavailable. The production of algae can be explained by three distinct mechanisms: photoautotrophic, mixotrophic, and heterotrophic mechanism of production. All these techniques follow the methods of natural growth. The production of photoautotrophic is generally autotrophic photosynthesis, the heterotrophic production needs glucose to promote growth, whereas some strains of some algae combine with heterotrophic assimilation of glucose and autotrophic photosynthesis in a mixotrophic procedure [40, 41].

7.6.1 Photoautotrophic Production

This kind of cultivation of microalgae are generally carried out when algae form carbohydrates by using inorganic carbon and sunlight by photosynthetic process. This method is the most wide-ranging method used for algae farming leading to the production of algal cells has a concentration of lipid ranges from 5 to 68%. This also depends upon the types of algal species being cultivated. Algae farming is generally done for the production of oil. Therefore, the basic benefits of utilizing this technique use CO_2 to meet the carbon necessity. Now, this technology is the only process which is economically and technologically viable for the production of algae biomass on a commercial large scale basis. The systems that have been organized are mainly based on closed photobioreactor and open pond technologies [42].

7.6.1.1 Open Pond Production Systems

The farming of algae in an open pond system has generally been adopted since the 1950s. The major algal species cultured in this system include spirulina, chlorella, and dunaliella.

- **Spirulina** is normally a filamentous cyanobacterium species which grows in alkaline lakes, soils, saline, freshwater, and brackish wastewater also. It converts nutrients into the cellular substances and releases O_2 by the process of photosynthesis. It needs H_2O along with Carbon, N, P, Fe, K, and some useful trace elements for survival and growth.

- The **microalgae (genus Chlorella)** normally synthesize extracellular and intracellular materials, including carbohydrates, proteins, and lipids. It has many benefits relative to higher plants, including higher % yield of biomass products and higher rate of growth without the use of arable land and can be used potentially for a source of nutrients from the wastewater effluents.
- The **microalgae** extracted from **genus Dunaliella** are cultivated because of their excellent antioxidant activity along with the enrichment of β-carotene, which have the capability to be accumulated during the conditions of stressful growth like elevated temperature and high salinity. The microalgae synthesize noticeable quantities of proteins, glycerol, and lipids.

The open pond method is the first microalgal synthesis technique and is still the most popular and widely used method adopted worldwide. This process generally operated with a low requirement of energy along with lower operational costs and initial investment. But, the controlling capability over the culture situations and pollution in these bioreactors are restricted. Their utilization of bioreactors for microalgae production has the capability of existing even in extreme climatic conditions [43, 44].

Raceway ponds are normally the most widely applied artificial system, which are usually a closed loop and have recirculation channels that are oval shaped (between 0.2 and 0.5 m deep). Proper circulation and mixing is necessary for stabilization of the productivity and growth of algae. The open raceway photobioreactors are considered as the most popular and widely applied microalgal culture systems used for the commercial scale of production. The raceway photobioreactors normally worked at depths of 15–30 cm with a flow rate of 15–30 cm s^{-1}. The raceway photo-bioreactors are usually the best culture systems for the production of microalgae. These systems provide the culture flows from the upper part of the bottom part in an inclined surface and is gathered in a retention tank, where it is pumped to the upper part of the system offering a higher rate of turbulence. In this system, there is a thin layer waterfall with slope changing from 1% to 3% and the depth of the culture system differs from some millimetres to 2 cm. The use of light source is highly efficient in these kinds of reactors, leading to the formation of higher productivity and biomass concentration. The raceway ponds are usually made in cement and concrete using white plastic as a liner. In the production cycle, the available nutrients and algae are inserted at the upper part of the paddlewheel. Further, the paddlewheel is circulated via a loop system. The continuous operation of the paddlewheel prevents sedimentation [45–47]. The open pond system is comparatively economical and does not require any land. Although production rates of algae biomass production are high, there are some discrepancies observed in the rate of production [48]. There are several classifications of algae as per their biochemical components, composition of the pigment life cycle, and ultra-structure (Figure 7.2).

7.6.1.2 Closed Photobioreactor Systems

The production of microalgae biomass by adopting the technology of closed photobioreactor is used to overcome the problems related to production in open pond systems. Like open pond systems, the closed system of photobioreactors supports the culture of single-species of microalgae for a longer time period without showing any risk towards the contamination of the ecosystem. The closed systems may be tubular, a column, or flat

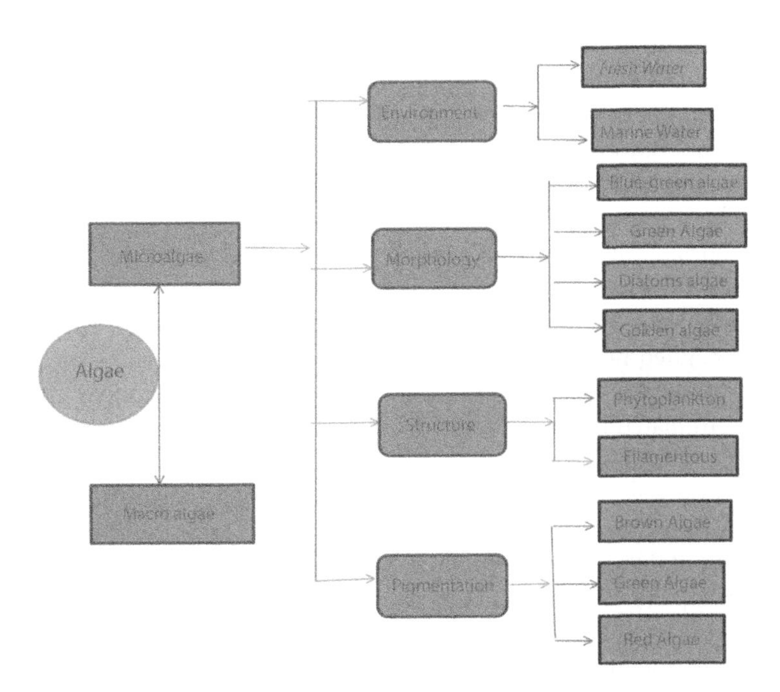

Figure 7.2 Classification of microalgae.

plate photobioreactor. These systems are suitable for sensitive strains. The management of the pollutant becomes easier due to closed configuration. Since in this process the highest cell biomass of microalgae is produced, the cost of production appreciably deceased. The costs in closed systems are comparatively higher than open pond systems.

A photobioreactor can be defined as a closed container used for phototrophic production where the energy source is provided through electric lights. The design of the rector should permit the efficient utilization of light energy and uniform illumination, while decreasing mutual shielding and offering a quick mass transfer of O_2 and CO_2. An ideal photobioreactor system consists of four phases:

- Microalgal cell biomass in solid phase
- Growth medium in liquid phase
- O_2 and CO_2 in gaseous phase
- Superimposed light radiation arena

Hence, for the design and development of an effective photobioreactor, knowledge regarding the complex interaction between the biomass yield and related environmental parameters (light transfer and fluid dynamics within the reactor system) is necessary. On the basis of the illuminated surface, the photobioreactor may be classified as flat plate, column, and tubular. Based on the flow of liquid, it is categorized as stirred type, airlift reactor, and bubble column reactors [49, 50].

An ideal photobioreactor should have to possess more transparent surface, less area of illumination, high rates of biomass transfer, and should attain a high growth rate of biomass. The design of a photobioreactor must be appropriate for farming of various kinds of microalgal species, which can prevent the fouling behaviour of the reactor.

The tubular group of photobioreactor traps the solar energy. These reactors are aligned horizontally, inclined, vertically, or of helical shape. The reactor tubes are normally of 0.1 m diameter or less. The algae cultures may be circulated through either an airlift system or mechanical pumping process. The former processes permit the exchange of O_2 and CO_2 between the aeration gas and in the liquid medium, which suggests a suitable mechanism for mixing. The process of mixing and agitation are treated as extremely vital to inspiring the exchange of gas within the tubes [51, 52].

The flat-plate photobioreactors are treated as the first kinds of closed systems. Now, it has given much attention to develop because of its huge surface area and greater densities of the photoautotrophic cells. These reactors are normally manufactured from suitable transparent materials having the capability to capture maximum solar energy and a very thin layer of compact culture flows across the flat plate. This may permit the absorbance of radiation within the first few millimetres of thickness. Because of less accumulation of DO (dissolved oxygen), a higher efficiency of photosynthetic activity is observed over the tubular reactor. It makes the flat-plate photo-bioreactors fit for the cultures of algae biomass.

The tubular photobioreactors are designed with less length of the tubes, being potentially dependent on the depletion of CO_2, accumulation of O_2 and variation of pH within the systems. Hence, the production of algae biomass requires the use of several reactor units on a large scale. But, the tubular photobioreactors are considered to be more appropriate for open-air mass cultures because of the huge surface exposed to sunlight [53, 54].

The column photobioreactors system provides the most effective mixing, highest rate of volumetric mass transfer, and best capability of regulating growth rate. These bioreactors are normally compact, low cost, portable, and easily operated. The vertical column photobioreactors are normally aerated by supplying air from its bottom part and are basically illuminated internally or via transparent walls. The performances of these photobioreactors are comparatively more than the tubular photobioreactors. Hence, all the groups of closed photobioreactors are focussed by researchers as compared to open raceway ponds because of its high rate of production of biomass with the significant potentiality of the production of biofuels along with useful by-products [55].

7.6.1.3 Hybrid Production Systems

The farming of microalgae in a hybrid system method consists of two stages and uses both photobioreactors and open ponds for various growth phases.

The first stage of farming is finished in a photobioreactor, where the continuous cell growth proceeds for the synthesis of biomass algae with less contaminated atmosphere under controlled conditions.

The second stage of farming is generally carried out in an open pond system and the culture medium is exposed to nutrient and environmental stresses. This causes an increase in the production of required lipid. This process currently is being used economically for the cultivation of algae.

The equipment set up and maintenance of open ponds is normally easier and it is one of the cheapest methods for cultivation of microalgae. The open ponds need less energy than photobioreactors, but are less effective than photobioreactors. There are some factors such as pollution, loss owing to evaporation, vacillation, temperature, incompetent mixing, and scarcity of light sources are some of the failures associated with the open pond systems [56, 57].

7.6.2 Heterotrophic Method Production

In this technique, the various algal species grow comfortably on a substrate of carbon, mainly glucose, and without any light source. This method can be achieved in a reactor having fewer surfaces to volume ratio. In this method because of the synthesis of high-density cells, a higher rate of controlling growth along with low harvest cost is achieved.

In this technique, the set up cost is less, but the energy consumption is comparatively more of the processes, which uses light energy due to photosynthetic activity and is exploited to form the source of carbon on which the microalgae survive and grow. The microalgae biomass produced in heterotrophic processes have a higher % of yield and the lipid contents in the cells is around 55%, relative to 15% in the case of autotrophic cells.

The heterotrophic method is now effectively and economically utilized for metabolites and production of algal biomass. In this procedure, the growth of microalgae does not depend on the light energy source, which permits for simple scale-up due to fewer surfaces to volume ratio.

In the heterotrophic method of microalgae production, the organic carbon consumed by microalgae is degraded or catabolized like bacteria. The microalgae are genetically able to metabolize the organic carbon as a source of carbon and energy. The carbon produced can metabolize enzymes present in the microalgae and replicates their primary mode of metabolic activity. It can store fixed carbon sources such as biopolymers (starch) and later on undergoes cleavage and is used in dark environments for cell division and growth of biomass products. Figure 7.3 indicates the scheme for complete production of biomass [58–60].

7.6.3 Mixotrophic Production

Some groups of microalgae have the ability to get nutrition through both heterotrophic and autotrophic processes. Hence, it indicates that the source of light is not the basic requirement for mixotrophs, as growth of cells generally happens via digestion of organic materials. This culture method reduces photo-inhibition and increases the rate of growth compared to heterotrophic and autotrophic cultures.

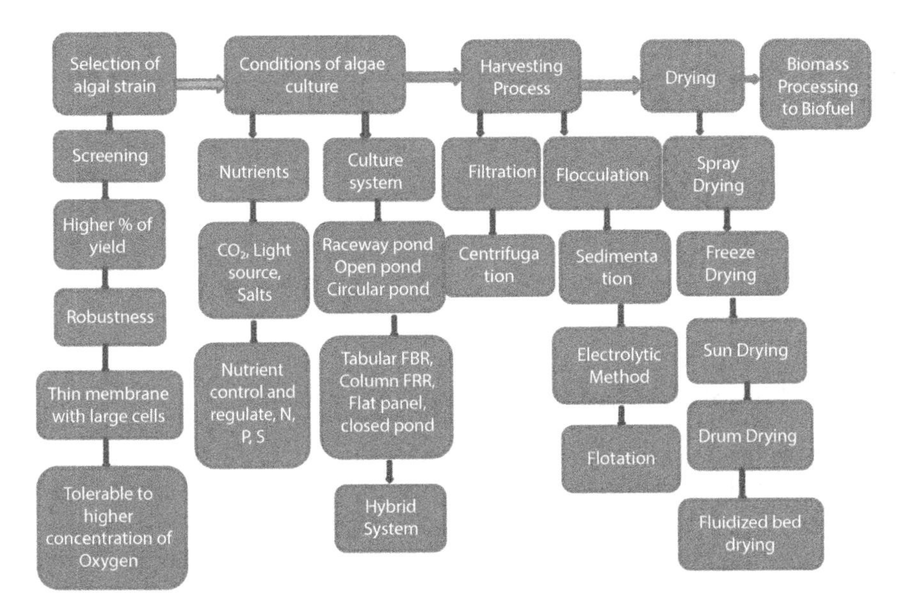

Figure 7.3 Schematic diagram for production of microalgae biomass systems.

The mixotrophic method of farming of microalgae uses both heterotrophic and photosynthetic elements, which retards biomass loss and again decreases the quantity of consumption of the organic substrate. Since it is the combination of heterotrophic and autotrophic methods, this process has the capability of photosynthesis and can ingest the organic substances. The capability of mixotrophs is to develop organic substrates, which means the cell growth is independent upon photosynthesis. Hence, the source of light energy is not only the limiting factor for the growth, because either substrates of organic carbon or light source stimulates the rate of growth [61, 62].

Examples: The microalgae showing the mixotrophic metabolism method for growth and existence are green algae Chlamydomonas reinhardtii and cyanobacteria Spirulina platensis [63].

Photosynthetic metabolism activity requires light for their growth and reproduction, while the aerobic respiration utilizes an organic carbon source. The rate of growth is affected by the media supplement with organic subtract (glucose) during dark and light phases with less loss of biomass product during the dark phase. The rate of growth of mixotrophic algae compared positively with the farming of photoautotrophic algae within a closed photobioreactor. The rate of growth is found to be more in open pond farming, but substantially lower than heterotrophic production.

The mixotrophic cultures retard the photoinhibition and develop the rate of growth over both heterotrophic and autotrophic cultures. Therefore, the formation of mixotrophic algae successfully permits the integration of both heterotrophic and photosynthetic mechanisms during the daytime cycle. This is the cause of the decrease in biomass loss during the dark respiration period and reduces the quantity of organic matter used during the growth phase of algae. These features conclude that the process of mixotrophic production is considered as a significant phase of the microalgae-to-biofuels conversion process [64, 65].

7.6.4 Photoheterotrophic Cultivation

In this method of cultivation, algae requires a source of light energy and carbon, which is obtained from the organic materials. It is the most widely used farming process for growth of algae, where algae utilize the organic carbon as both the carbon and energy sources. In photoheterotrophic cultivation, algae needs light for utilization of organic substances as the source of carbon. The major distinction between photoheterotrophic and mixotrophic farming is that the former needed light as the source of energy, while the latter used organic substances for this purpose.

Algal growth by using the photoautotrophic method normally depends on nutrients, mainly N, C, P, and micronutrients. The water soluble form of these nutrients can be effectively used for culture of microalgae [66, 67].

Example: Algae can use food supplements such as DHA or cosmetics and are always farmed in the synthetic medium. The domestic, industrial, and animal wastewater effluents are used as the medium of cultivation [68].

Presently, the largest production of biofuels from sources of algae is biodiesel and the synthesis of biofuel based on biomass algae is highly beneficial to our environment because the biomass growth takes CO_2 from the atmosphere and the combustion of these kinds of fuels generates fewer amounts of greenhouse gases compared to petroleum fuels.

Biodiesel is normally manufactured due to the chemical reaction of lipids with an alcohol forming fatty acid esters. The chlorophyll-A and neutral lipids synthesized by microalgae serve as an excellent source of lipid during the production of biodiesel. Hence, for extracting lipids or algal oil in conventional method, the steps involved are drying the biomass product of microalgae followed by extraction through organic solvent. By using modern technology, we get directly wet algae, but this process is still in an undeveloped stage.

Although algae is a versatile biota in the environment, the commercial production of algae based biofuel in the present scenario is not economically viable because of many technical barriers. It is essential but very problematic for the development of economical methods of harvesting algae and extraction of lipid, therefore the biofuel technology utilizing algae biomass requires modification [69, 70].

7.7 Impact of Microalgae on the Environment

The impact of microalgae has several positive upshots on the environment. Some of them are discussed here.

- Microalgae lipid productivity is much more viable compared to the traditional food crops. It needs less area for cultivation compared to food crops, therefore saving our agricultural land. The production cost of algae fuel is comparatively higher than the stock resource fossil fuels. In the present situation, its development is highly essential to protect our globe from the negative impact of fossil fuel combustion and to save our valuable stock resources for future generations.
- Hence, the cost of production of biofuel from microalgae biomass requires analysis and modification to decrease the initial capital cost and operating cost for the commercial scale of production of microalgae. For farming of microalgae, the present technique can be developed partly via simultaneous water treatment and production of algal biomass together in a hybrid technology in a mixotrophic environment. The microalgae grows by consuming the solar energy by the process of photosynthesis and converts the solar energy to chemical energy. This actually promotes the growth of microalgae [71, 72].

Example: Microalgae needed less water, minimum intensity of sunlight, and simple nutrients (N, P, K, S) for their growth.

- Due to the tolerance capability of the microalgae in drastic environmental conditions, it can be cultivated effectively in any local adverse environment, which is not appropriate for the cultivation of feedstock such as palm oil, sunflower, and soybean for production of biodiesel in the existing technology. The conventional use fuels in combustion, produce a lot of contaminants such as CO_2, CO, NO_x, SO_x, and some other toxic gases along with particulate matter (PM). However, the biofuel obtained from algae biomass is free from sulphur and no PM, SO_x, NO_x, hydrocarbons, or CO are emitted during its combustion [73].

- Another benefit of the microalgae farming for the production of biofuel is also partly responsible for the removal of CO_2 from the flue gases emitted from industrial activities and coal fired power plants through bio-fixation. The cultivation of microalgae in wastewater effluents or sludge water is highly beneficial. The microalgae consume NH_4^+, PO_4^{3-}, and NO_3^- as the source of nutrients from the wastewater and grow effectively. This technique can decrease the requirement of fresh water for the cultivation of algae and simultaneously remediate the toxic wastewater.
- In addition to that, another benefit of using microalgae is that, after extraction of oil from the biomass, the residue obtained also possesses a high value of N:P ratio and can be effectively used as fertilizers. The biomass residuals after the production of biofuel can also be processed in different ways and some valuable products such as natural dyes, antioxidants, protein, carbohydrates, and pigments are derived which can be used as cosmeceutical products or can be made into nutritional supplements.
- The cultivation of algae for the production of biofuel is beneficial because it produces biomass products at a rapid rate and is more eco-friendly towards the environment. In comparison to fossil fuels, the cost of production is much higher in the present technologies. The cultivation of algae in wastewater and NPK enriched water sources is effective because algae get sufficient nutrients from the medium. Presently, a number of scenarios were reviewed in order to optimize the negative impacts towards the environment where the cultivation of microalgae in a nutrient rich medium is effective. It was identified that the energy generated by biomass of algae in the open pond systems was found to be higher than the energy spent by a PBR system, but both of these techniques are not economically viable [74, 75].
- In open pond technology, the emission of greenhouse gases (GHG) is much less than PBR systems because of anaerobic digestion of the residual biomass products in the open pond system. The common facts in these two systems are:

- The production of renewable diesel via transesterification, pyrolysis, and hydrothermal liquefaction
- The generation of biogas via anaerobic digestion without or with hydrothermal pre-treatment

In both the methods, the consumption of energy was basically for bioenergy conversion and fertilizer production [76].

7.8 Advantages of Utilizing Microalgae Biomass for Biofuels

Microalgae cultivation for the purpose of biofuel production is treated as an extremely promising approach due to several advantages.

- The cultivation of microalgae does not need any arable land
- It can be effectively cultivated in unused lands, degraded soils, desert areas, and even in wastewater effluents
- Therefore, the fertile lands can be used for agricultural purposes and no interference in food production occurs
- The cultivation of microalgae is not competing with normal agriculture for food based crops
- The algae can be cultivated in any season. It does not depend upon the climatic conditions. It can grow effectively even in harsh climatic situations.
- The biomass productivity is much higher than the crops and quickly accumulates the lipids in 15-50% of its dry mass in most of the species
- Although it can grow in an aqueous medium in addition to any land surface, it can grow effectively in less quantity of water than most of the terrestrial plant species
- The residues of it can be utilized for production of some other valuable products along with fertilizers
- Since it can be grown in wastewater and sewage water by consuming nutrients from it, in addition to biomass production, it partly remediates wastewater
- Microalgae generate more than 50% of the oxygen present in the atmosphere
- The cultivation of microalgae does not need any kinds of fertilizers, pesticides, and herbicides for their growth

- The microalgae gets nutrients from agro-industrial wastes and wastewater effluents and has no need for the addition of extra nutrients for its growth
- The microalgae can effectively fix atmospheric carbon and the residues obtained from industrial activities through the process of photosynthesis
- The residual products after biofuel production can also be processed in different ways and forms into useful products such as carbohydrates, proteins, and carotenoids. The residues can be successfully utilized as fertilizers or in some cases food. Its fermentation can also produce ethanol [77–79].

7.9 Conclusion

The existing technology for production of biofuel from microalgae is a viable renewable energy resource, but its use on a commercial basis to replace the fossil fuel is not viable due to so many demerits. Microalgae are considered interesting and attractive resources of feedstock for the production of biofuel, including biogas, biodiesel, methane, bio-hydrogen, bioethanol, and bio-methanol along with some valuable materials. Now, in the present situation, the commercialization of biofuel is highly important to save our environment and biological community from drastic degradation in the quality of air. The biofuel is eco-friendly, renewable, sustainable, and biodegradable in nature. There are so many advantages of algae such as high rate of growth, easily cultivation without any land, and substantially high lipid content. The higher rate of yield of both biomass and lipids can be achieved due to the combined effect of algae and some suitable co-products, if decisively exploited, and decreases the cost of production of biofuel from microalgae. The continuous modification of technologies can optimize the production of microalgae, processing of biomass, and oil extraction and contributes significantly to achieving the goal of biofuel production from microalgae. Still, improvement in technology is needed to overcome the disadvantages and commercialization of this process.

References

1. Gupta, P. L., Lee, S.-M., & Choi, H.-J. A mini review: photobioreactors for large scale algal cultivation. *World Journal of Microbiology and Biotechnology*, 31(9), 1409–1417 (2015). doi:10.1007/s11274-015-1892-4

2. Onar, O. C., & Khaligh, A. Energy Sources. Alternative Energy in Power Electronics, 81–154 (2015). doi:10.1016/b978-0-12-416714-8.00002-0
3. Abas, N., Kalair, A., & Khan, N. Review of fossil fuels and future energy technologies. *Futures*, 69, 31–49 (2015). doi:10.1016/j.futures.2015.03.003
4. Alalwana, H. A., Alminshid A. H. and Aljaafaric, H. A.S. Promising evolution of biofuel generations. Subject review. *Renewable Energy Focus* 28, 127-139(2019). https://doi.org/10.1016/j.ref.2018.12.006
5. Alam, F., Mobin, S. and Chowdhury, H. Third generation biofuel from Algae. *Procedia Engineering* 105, 763 -768 (2015).
6. Sadatshojaei, E., Wood, D.A., Mowla, D. Third Generation of Biofuels Exploiting Microalgae. In: Inamuddin, Asiri A. (eds) Sustainable Green Chemical Processes and their Allied Applications. *Nanotechnology in the Life Sciences. Springer, Cham.* 575-588 (2020). https://doi.org/10.1007/978-3-030-42284-4_21
7. Mohr, A.,& Raman, S. Lessons from first generation biofuels and implications for the sustainability appraisal of second generation biofuels. *Energy Policy*, 63,114–122. (2013) doi:10.1016/j.enpol.2013.08.033
8. Enamala, M. K., Enamala, S., Chavali, M., Donepudi, J., Yadavalli, R., Kolapalli, B., Kuppam, C. Production of biofuels from microalgae - A review on cultivation, harvesting, lipid extraction, and numerous applications of microalgae. *Renewable and Sustainable Energy Reviews*, 94, 49–68 (2018). doi:10.1016/j.rser.2018.05.012
9. Yin, Z., Zhu, L., Li, S., Hu, T., Chu, R., Mo, F., Li, B. A comprehensive review on cultivation and harvesting of microalgae for biodiesel production: environmental pollution control and future directions. Bioresource Technology, 301, 1-19 (2020). 122804. doi:10.1016/j.biortech.2020.122804
10. Ganesan, R., Manigandan, S., Samuel, M. S., Shanmuganathan, R., Brindhadevi, K., Lan Chi, N. T. Pugazhendhi, A. A review on prospective production of biofuel from microalgae. *Biotechnology Reports*, 27, 1-13 (2020),e00509.
11. Thiruvenkadam, S., Izhar, S., Yoshida, H., Danquah, M. K., & Harun, R. Process application of Subcritical Water Extraction (SWE) for algal bio-products and biofuels production. *Applied Energy*, 154, 815–828 (2015).
12. Hannon, M., Gimpel, J., Tran, M., Rasala, B. and Mayfield, S. Biofuels from algae: challenges and potential, *Biofuels.* 1(5): 763–784 (2010).
13. Khan, S., Siddique, R., Sajjad, W., Nabi, G., Hayat, K. M., Duan, P., & Yao, L. Biodiesel Production From Algae to Overcome the Energy Crisis. *HAYATI Journal of Biosciences*, 24(4), 163–167 (2017). doi:10.1016/j.hjb.2017.10.003
14. Leite, G. B., Abdelaziz, A. E. M., & Hallenbeck, P. C. Algal biofuels: Challenges and opportunities. *Bioresource Technology*, 145, 134–141 (2013).
15. Osanai, T., Park, Y.-I., & Nakamura, Y. Editorial: Biotechnology of Microalgae, Based on Molecular Biology and Biochemistry of Eukaryotic Algae and Cyanobacteria. *Frontiers in Microbiology*, 8, 118-121 (2017).

16. Demoulin, C. F., Lara, Y. J., Cornet, L., François, C., Baurain, D., Wilmotte, A., & Javaux, E. J. Cyanobacteria evolution: Insight from the fossil record. *Free Radical Biology and Medicine.* 140, 206-223 (2019).
17. Bhateria, R., & Dhaka, R. Algae as biofuel. *Biofuels*, 5(6), 607–631 (2014). doi:10.1080/17597269.2014.1003701
18. Saad, A., and Atia, A., Review on Freshwater Blue-Green Algae (Cyanobacteria): Occurrence, Classification and Toxicology, *Biosciences Biotechnology Research Asia.* 11(3), 1319-1325 (2014).
19. Sathasivam, R., Radhakrishnan, R., Hashem, A., & Abd_Allah, E. F. Microalgae metabolites: A rich source for food and medicine. *Saudi Journal of Biological Sciences.* 26(4), 709-722 (2019). doi:10.1016/j.sjbs.2017.11.003
20. Perrine, Z., Negi, S., & Sayre, R. T. Optimization of photosynthetic light energy utilization by microalgae. *Algal Research*, 1(2), 134–142 (2012).
21. Abdel-Raouf, N., Al-Homaidan, A. A., & Ibraheem, I. B. M. Microalgae and wastewater treatment. *Saudi Journal of Biological Sciences*, 19(3), 257–275 (2012). doi:10.1016/j.sjbs.2012.04.005
22. Alam, F, Datea, A., Rasjidina, R., Mobinb, S., Moriaa, H., Biofuel from algae- Is it a viable alternative? *Procedia Engineering* 49, 221-227 (2012).
23. Deshmukh, S., Kumar, R., & Bala, K., Microalgae biodiesel: A review on oil extraction, fatty acid composition, properties and effect on engine performance and emissions. Fuel Processing Technology, 191, 232–247 (2019).
24. Xue, Z., Yu, Y., Yu, W., Gao, X., Zhang, Y., & Kou, X., Development Prospect and Preparation Technology of Edible Oil From Microalgae. *Frontiers in Marine Science*, 7.402-416 (2020). doi:10.3389/fmars.2020.00402
25. Hassan, M. H., & Kalam, M. A. An Overview of Biofuel as a Renewable Energy Source: Development and Challenges. *Procedia Engineering*,56, 39–53 (2013). doi:10.1016/j.proeng.2013.03.087
26. Jalilian, N., Najafpour, G. D., & Khajouei, M., Macro and Micro Algae in Pollution Control and Biofuel Production – A Review. *ChemBioEng Reviews.* 7, 1-17 (2020). doi:10.1002/cben.201900014
27. Hoyer, J., Cotta, F., Diete, A., & Großmann, J. Bioenergy from Microalgae - Vision or Reality? *ChemBioEng Reviews*, 5(4), 207–216 (2018).
28. Correa, D. F., Beyer, H. L., Fargione, J. E., Hill, J. D., Possingham, H. P., Thomas-Hall, S. R., & Schenk, P. M. Towards the implementation of sustainable biofuel production systems. *Renewable and Sustainable Energy Reviews*, 107, 250-263 (2019). https://doi.org/10.1016/j.rser.2019.03.005
29. Rodionova, M. V., Poudyal, R. S., Tiwari, I., Voloshin, R. A., Zharmukhamedov, S. K., Nam, H. G., Allakhverdiev, S. I., Biofuel production: Challenges and opportunities. *International Journal of Hydrogen Energy*, 42(12), 8450–8461 (2017). doi:10.1016/j.ijhydene.2016.11.125
30. Shuba, E. S., & Kifle, D., Microalgae to biofuels: “Promising” alternative and renewable energy, review. *Renewable and Sustainable Energy Reviews*, 81,743–755 (2018). doi:10.1016/j.rser.2017.08.042

31. Raheem, A., Prinsen, P., Vuppaladadiyam, A. K., Zhao, M., & Luque, R. A review on sustainable microalgae based biofuel and bioenergy production: Recent developments. Journal of Cleaner Production, 181, 42–59 (2018).
32. Abo, B. O., Odey, E. A., Bakayoko, M., & Kalakodio, L. Microalgae to biofuels production: a review on cultivation, application and renewable energy. Reviews on Environmental Health, 34(1), 1-9 (2019) doi:10.1515/reveh-2018-0052
33. Saad, M. G., Dosoky, N. S., Zoromba, M. S., & Shafik, H. M., Algal Biofuels: Current Status and Key Challenges. *Energies*, 12(10), 1920-1941 (2019).
34. Voloshin, R. A., Rodionova, M. V., Zharmukhamedov, S. K., Nejat Veziroglu, T., & Allakhverdiev, S. I. Review: Biofuel production from plant and algal biomass. *International Journal of Hydrogen Energy*, 41(39), 17257–17273 (2016). doi:10.1016/j.ijhydene.2016.07.084
35. Makareviciene, V., Skorupskaite, V. & Andruleviciute, V. Biodiesel fuel from microalgae-promising alternative fuel for the future: a review. *Rev Environ Sci Biotechnol* 12, 119–130 (2013). https://doi.org/10.1007/s11157-013-9312-4
36. Singh, N. K., & Dhar, D. W., Microalgae as second generation biofuel. A review. *Agronomy for Sustainable Development*, 31(4), 605–629 (2011). doi:10.1007/s13593-011-0018-0
37. Brennan, L., & Owende, P. Biofuels from microalgae—A review of technologies for production, processing, and extractions of biofuels and co-products. *Renewable and Sustainable Energy Reviews*, 14(2), 557–577 (2010). doi:10.1016/j.rser.2009.10.009
38. Dębowski, M., Zieliński, M., & Dudek, M. Concept of a Technological System for Microalgae Biomass Production with the Use of Effluents from Fermentation Tanks. Energy Procedia, 105, 681–687 (2017). doi:10.1016/j.egypro.2017.03.375
39. Falciatore, A., Jaubert, M., Bouly, J.-P., Bailleul, B., & Mock, T. Diatom Molecular Research Comes of Age: Model Species for Studying Phytoplankton Biology and Diversity. *The Plant Cell*, 32, 547–572 (2020).
40. Show, P. L., Tan, J. S., Lee, S. Y., Chew, K. W., Lam, M. K., Lim, J. W., & Ho, S.-H.). A Review on Microalgae Cultivation and Harvesting, and Their Biomass Extraction Processing Using Ionic Liquids. *Bioengineered*.11(1),116-129(2020) doi:10.1080/21655979.2020.1711626
41. Dineshkuma, R., Narendran, R. & Sampathkumar, P., Cultivation and harvesting of micro-algae for bio-fuel Production – A review, *Indian Journal of Geo Marine Sciences* 46 (09), 1731-1742 (2017).
42. Noreña-Caro, D., & Benton, M. G., Cyanobacteria as photoautotrophic biofactories of high-value chemicals. *Journal of CO_2 Utilization*, 28, 335–366 (2018).
43. Sreekumar, N., Giri Nandagopal, M. S., Vasudevan, A., Antony, R., & Selvaraju, N. Marine microalgal culturing in open pond systems for biodiesel production—Critical parameters. *Journal of Renewable and Sustainable Energy*, 8(2), 1-19 (2016). 023105. doi:10.1063/1.4945574

44. Kannan, D. C., & Venkat, D. An open outdoor algal growth system of improved productivity for biofuel production. Journal of Chemical Technology & Biotechnology. 94(1), 222-235 (2018). doi:10.1002/jctb.5768
45. Sudhakar, K., Premalatha, M., & Rajesh, M., Large-scale open pond algae biomass yield analysis in India: a case study. *International Journal of Sustainable Energy*, 33(2), 304–315 (2014). doi:10.1080/14786451.2012.710617
46. Rayen, F., Behnam, T., & Dominique, P. Optimization of a raceway pond system for wastewater treatment: a review. *Critical Reviews in Biotechnology*, 39(3)1–14 (2019). doi:10.1080/07388551.2019.1571007
47. Narala, R. R., Garg, S., Sharma, K. K., Thomas-Hall, S. R., Deme, M., Li, Y., & Schenk, P. M. (2016). Comparison of Microalgae Cultivation in Photobioreactor, Open Raceway Pond, and a Two-Stage Hybrid System. Frontiers in Energy *Research*, 4, 29-40. doi:10.3389/fenrg.2016.00029
48. Richardson, J. W., Johnson, M. D., & Outlaw, J. L., Economic comparison of open pond raceways to photo bio-reactors for profitable production of algae for transportation fuels in the Southwest. *Algal Research*, 1(1), 93–100 (2012). doi:10.1016/j.algal.2012.04.001
49. Zhou, X., Yuan, S., Chen, R., & Ochieng, R. M., Sustainable production of energy from microalgae: Review of culturing systems, economics, and modelling. *Journal of Renewable and Sustainable Energy*, 7(1),1-29 (2015).
50. Singh, R. N., & Sharma, S., Development of suitable photobioreactor for algae production – A review. *Renewable and Sustainable Energy Reviews*, 16(4), 2347–2353 (2012). doi:10.1016/j.rser.2012.01.026
51. Huang, Q., Jiang, F., Wang, L., & Yang, C. (2017). Design of Photobioreactors for Mass Cultivation of Photosynthetic Organisms. Engineering, 3(3), 318–329. doi:10.1016/j.eng.2017.03.020
52. Mata, T. M., Cameira, M., Marques, F., Santos, E., Badenes, S., Costa, L., Martins, A. A., Carbon footprint of microalgae production in photobioreactor. Energy Procedia, 153, 432–437 (2018). doi:10.1016/j.egypro.2018.10.039
53. Sierra, E., Acién, F. G., Fernández, J. M., García, J. L., González, C., & Molina, E., Characterization of a flat plate photobioreactor for the production of microalgae. *Chemical Engineering Journ*al, 138(1-3),136–147 (2008).
54. Fernández, I., Acién, F. G., Berenguel, M., & Guzmán, J. L., First Principles Model of a Tubular Photobioreactor for Microalgal Production. Industrial & Engineering Chemistry Research, 53(27), 11121–11136 (2014). doi:10.1021/ie501438r
55. Xu, L., Weathers, P. J., Xiong, X.-R., & Liu, C.-Z. Microalgal bioreactors: Challenges and opportunities. *Engineering in Life Sciences*, 9(3), 178–189 (2009).
56. Deprá, M. C., Mérida, L. G. R., de Menezes, C. R., Zepka, L. Q., & Jacob-Lopes, E., A new hybrid photobioreactor design for microalgae culture. *Chemical Engineering Research and Design*. 144, 1-10 (2019). doi:10.1016/j.cherd.2019.01.023

57. Pal, P., Chew, K. W., Yen, H.-W., Lim, J. W., Lam, M. K., & Show, P. L., Cultivation of Oily Microalgae for the Production of Third-Generation Biofuels. *Sustainability*, 11(19), 5424-5439 (2019). doi:10.3390/su11195424
58. Shamzi Mohamed, M., Zee Wei, L., & B. Ariff, A. Heterotrophic Cultivation of Microalgae for Production of Biodiesel. *Recent Patents on Biotechnology*, 5(2), 95–107 (2011). doi:10.2174/187220811796365699
59. Morales-Sánchez, D., Martinez-Rodriguez, O. A., & Martinez, A. Heterotrophic cultivation of microalgae: production of metabolites of commercial interest. *Journal of Chemical Technology & Biotechnology*, 92(5), 925–936 (2016).
60. Ende, S.S.W., Noke, A. Heterotrophic microalgae production on food waste and by-products. *J Appl Phycol* 31, 1565–1571 (2019). https://doi.org/10.1007/s10811-018-1697-6
61. Lowrey, J., Brooks, M.S. & McGinn, P.J. Heterotrophic and mixotrophic cultivation of microalgae for biodiesel production in agricultural wastewaters and associated challenges—a critical review. *J Appl Phycol* 27, 1485–1498 (2015).
62. Zhan, J., Rong, J., & Wang, Q. Mixotrophic cultivation, a preferable microalgae cultivation mode for biomass/bioenergy production, and bioremediation, advances and prospect. *International Journal of Hydrogen Energy*, 42(12), 8505–8517 (2017). doi:10.1016/j.ijhydene.2016.12.021
63. Li, Y.-R., Tsai, W.-T., Hsu, Y.-C., Xie, M.-Z., & Chen, J.-J. ,Comparison of Autotrophic and Mixotrophic Cultivation of Green Microalgal for Biodiesel Production. *Energy Procedia*, 52, 371–376 (2014). doi:10.1016/j.egypro.2014.07.088
64. Wang, J., Yang, H., & Wang, F. (2014). Mixotrophic Cultivation of Microalgae for Biodiesel Production: Status and Prospects. Applied Biochemistry and Biotechnology, 172(7), 3307–3329. doi:10.1007/s12010-014-0729-1
65. Pereira, M. I. B., Chagas, B. M. E., Sassi, R., Medeiros, G. F., Aguiar, E. M., Borba, L. H. F., Rangel, A. H. N., Mixotrophic cultivation of Spirulina platensis in dairy wastewater: Effects on the production of biomass, biochemical composition and antioxidant capacity. *PLOS ONE*, 14(10), 1-17 (2019). e0224294.
66. Chen, C.-Y., Yeh, K.-L., Aisyah, R., Lee, D.-J., & Chang, J.-S., Cultivation, photobioreactor design and harvesting of microalgae for biodiesel production: A critical review. *Bioresource Technology*, 102(1), 71–81 (2011).
67. Liu,C.-H., Chang, C.-Y., Liao, Q., Zhu, X., & Chang, J.-S, Photoheterotrophic growth of Chlorella vulgaris ESP6 on organic acids from dark hydrogen fermentation effluents. *Bioresource Technology*, 145, 331–336 (2013).
68. Nur, M.M.A., Buma, A.G.J. Opportunities and Challenges of Microalgal Cultivation on Wastewater, with Special Focus on Palm Oil Mill Effluent and the Production of High Value Compounds. Waste Biomass Valor 10, 2079–2097 (2019). https://doi.org/10.1007/s12649-018-0256-3

69. Patel, A., Matsakas, L., Rova, U. *et al.* Heterotrophic cultivation of Auxenochlorella protothecoides using forest biomass as a feedstock for sustainable biodiesel production. *Biotechnol Biofuels* 11, 169-184 (2018). https://doi.org/10.1186/s13068-018-1173-1
70. Benavente-Valdés, J. R., Aguilar, C., Contreras-Esquivel, J. C., Méndez-Zavala, A., & Montañez, J., Strategies to enhance the production of photosynthetic pigments and lipids in chlorophycae species. *Biotechnology Reports*, 10, 117–125 (2016). doi:10.1016/j.btre.2016.04.001
71. Usher, P. K., Ross, A. B., Camargo-Valero, M. A., Tomlin, A. S., & Gale, W. F., An overview of the potential environmental impacts of large-scale microalgae cultivation. *Biofuels*, 5(3), 331–349 (2014). doi:10.1080/17597269.2014.913925
72. Han, P., Lu, Q., Fan, L., & Zhou, W. A Review on the Use of Microalgae for Sustainable Aquaculture. *Applied Sciences*, 9(11), 2377-2396 (2019).
73. Bhola, V., Swalaha, F., Ranjith Kumar, R. *et al.* Overview of the potential of microalgae for CO2 sequestration. *Int. J. Environ. Sci. Technol.* 11, 2103–2118 (2014). https://doi.org/10.1007/s13762-013-0487-6
74. Show, P., Tang, M., Nagarajan, D., Ling, T., Ooi, C.-W., & Chang, J.-S. A Holistic Approach to Managing Microalgae for Biofuel Applications. International Journal of Molecular Sciences, 18(1), 215-249 (2017). doi:10.3390/ijms18010215
75. Sharma PK, Saharia M, Srivastava R, Kumar S and Sahoo L., Tailoring microalgae for efficient biofuel production, *Front. Mar. Sci.* 5:382-437(2018).
76. Han, S.-F., Jin, W.-B., Tu, R.-J., & Wu, W.-M., Biofuel production from microalgae as feedstock: current status and potential. *Critical Reviews in Biotechnology*, 35(2), 255–268 (2014). doi:10.3109/07388551.2013.835301
77. Simas-Rodrigues, C., Villela, H. D. M., Martins, A. P., Marques, L. G., Colepicolo, P., & Tonon, A. P., Microalgae for economic applications: advantages and perspectives for bioethanol. *Journal of Experimental Botany*, 66(14), 4097–4108 (2015). doi:10.1093/jxb/erv130
78. Kröger, M., & Müller-Langer, F., Review on possible algal-biofuel production processes. *Biofuels*, 3(3), 333–349 (2012). doi:10.4155/bfs.12.14
79. Koyande, A. K., Show, P.-L., Guo, R., Tang, B., Ogino, C., & Chang, J.-S., Bio-processing of algal bio-refinery: review on current advances and future perspectives. *Bioengineered*, 10(1), 574–592 (2019). doi:10.1080/21655979.2019.1679697

8

Agrowaste Lignin as Source of High Calorific Fuel and Fuel Additive

Harit Jha* **and Neha Namdeo**

Department of Biotechnology, Guru Ghasidas Vishwavidyalaya, Bilaspur, Chhattisgarh, India

Abstract

Lignin is the second most abundant biopolymer found on earth, which accounts for about 30-35% of organic carbon within the environment. Lignin is biodegradable and provides rigidity and strength to the stems as well as maintains structural integrity of the cell wall. Lignin is used as a source of many valuable chemicals, fuels, additives, etc. It also has many applications like biopolymer formation (biofilm, resins, foam), antioxidants, plant growth promoters, metal ion chelating agents, surfactant synthesis of nano fibers/carbon fibers/tube aero gel/hydrogel, composites, an adsorbent agent, an emulsifier, a chelating agent, a dispersant reagent in cement, and as gypsum beads, etc. Agrowaste lignin also serves as a very important source for production of fuels like bioethanol, bio-oil, and syngas. It is one of the waste products obtained during bioethanol production, though some of it is used for production of char. Thus, it is an important and easily available agrowaste based biopolymer used for production of fuel.

Keywords: Agrowaste, biopolymer, lignin, fuel

8.1 Agrowaste

The need to convert waste to wealth triggered many researchers to develop methods for converting agrowaste into useful products. Traditionally, agrowaste materials were used as lost circulation materials and a mixture of two or more agrowaste materials prevents lost circulation better than

**Corresponding author*: harit74@yahoo.co.in

Prakash Kumar Sarangi (ed.) Biorefinery Production of Fuels and Platform Chemicals, (179–202) © 2023 Scrivener Publishing LLC

when used alone. Agrowaste materials are inexpensive, ubiquitous, and are easily available. The potential of agrowastes to be transformed into a range of utilizable products is indisputable as their usage is found in various fields such as building construction, woodwork, civil engineering, etc. Most of its technical applications have been in the field of lost circulation prevention [1]. Agrowaste plays a crucial role in sustainable energy production. Mostly a large amount of waste is produced by the agricultural sector in the form of paddy straw, wheat straw, and corn cob, which helps to meet the increasing demand of fuel in an economical manner [2, 3]. The agricultural sector is the largest in world's economy and contributes 1/3rd of global GDP, 26.81% of global employment, and 38.14% of land area. In India, around 600 million tons of agricultural residue is produced per year [4]. These include animal waste, field waste, and agro-industrial waste. Primary and secondary crop wastes mainly include straws, cobs, stalks, etc. which are rich in energy which can meet the energy demands [3, 5].

8.2 Lignin

Lignin, after cellulose, is the second most abundant biopolymer or aromatic biopolymer found on earth which accounts for about 30-35% of organic carbon within the environment [6] and is also one of the three most abundant renewable resources including cellulose and natural oils [7]. The deposition of lignin in the cell wall of plant plays an important role in plant physiology and development: (i) by providing mechanical support to plant organs, allowing large size and upright growth; (ii) by providing rigidity and strength to the cells; (iii) by allowing water and solute transportation in the vascular system due to its hydrophobicity and mechanical resistance; and (iv) associated with protection against pathogens [8–10]. There is different lignin content in various types of wood, i.e. 15–25% in grasses, 19–28% in hardwoods, and 24–33% in softwoods, respectively [11].

There has been an evolutionary adaptation of plants to synthesize lignin from the marine environment to land. Lignin is responsible for the rigidity and strength of the stem as well as the structural integrity of the cell wall [6]. The only large volume renewable aromatic foodstock found in most of the terrestrial plants is lignin [12]. The yearly production of lignin is more than 70 million tons per year. Industrial lignins are produced from sulfite and kraft pulping processes in the paper and pulp industries, so-called black liquor. Most of the lignin produced from pulp industries is burned as a low-value fuel for energy supplement, leading to waste of important resources and environmental pollution, whereas only less than 2% of the lignin produced from these industries were used for value added stuffs [11]. Various functional groups and the polymeric nature make it flexible for different industrial applications [13].

8.2.1 Structure of Lignin

Lignin is an irregular three-dimensional, amorphous, and highly branched phenolic polymer and accounts for 15–30% by weight of lignocellulosic biomass (LCBM) [11]. The structure of lignin is comprised of the combinatorial oxidative coupling of three basic phenylpropanoid monomers, p-hydroxyphenyl (H), guaiacyl (G), and syringyl (S) units, derived from *p*-coumaryl, coniferyl, and sinapyl alcoholic precursors, respectively (Figure 8.1) [8, 9, 13].

Some other monomers were also known to participate in polymerization of lignin, which includes aldehydes and hydroxycinnamic acids, as well as sinapyl acetates or coumarates and coniferyl [15]. After the synthesis of lignin monomers, they are transported to the cell wall where they are polymerized in a combinatorial fashion mediated by peroxidases, using free radical coupling mechanisms giving rise to numerous structures within the lignin polymers [8]. Depending on the environmental conditions, tissues, and cell location, the relative proportion of lignin monomers varies between plants. The plant produces lignin with a specific composition which depends on the precursors that are deposited in the lignifying zone, thus providing high chemical flexibility to the lignin molecule [16, 17]. The composition and content of lignin may differ amongst various environmental conditions or plant growth stages, taxa, cell type, and

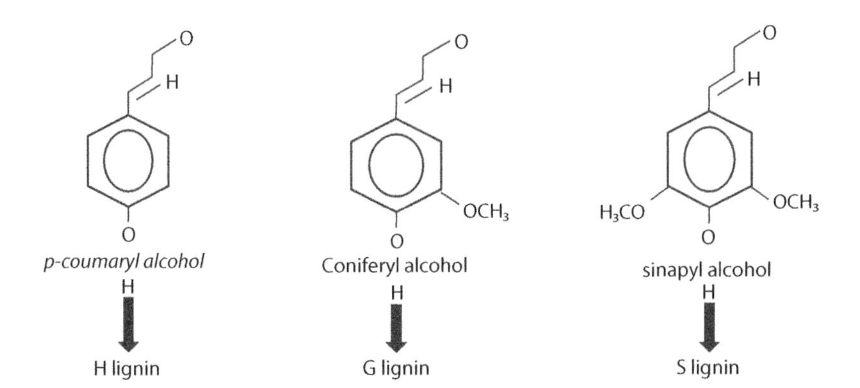

Figure 8.1 Monomers of lignin derived from alcoholic precursors.

individual cell wall layers [10, 16]. Carbonyl, carboxyl, methoxyl, and hydroxyl are the functional groups present in lignin which are linked to aromatic or aliphatic moieties, having different amounts, leading to various structures and composition of lignin [18]. Many linkages either in C-C or C-O type with different abundances are formed in the coupling reactions involved in biosynthesis of lignin, including aryglycerol-β-ether dimer (β-O-4, 45–50%), biphenyl/dibenzodioxocin (5–5′, 18–25%), pino/resinol (β-5, 9–12%), diphenylethane (β-1, 7–10%), aryglycerol-α-ether dimer (α-O-4, 6–8%), phenylcoumaran (β-β', 0–3%), siaryl ether (4-O-5, 4–8%), and spirodienon [11, 14]. The determination of chemical structure and spectral imaging of lignin is done using new methods such as confocal Raman scattering microscopy [19] and time-of-flight secondary ion mass spectrometry [19, 20]. These methods are available only in biological laboratories and have not been employed by chemical scientific groups. Until now, the structural elucidation and compositional characteristics of lignin depended on these processes for the isolation and degradation of lignin from LCBM and methods used for the characterization of the analogous products [21, 22]. Structure, chemical composition, and bonding between lignin in biomass is shown in Table 8.1.

Table 8.1 Structure, chemical composition, and bonding between lignin in biomass.

Characters	Lignin
Polymer	G lignin, GS lignin, GSH lignin
Subcomponent	Guaiacyl (G), syringyl (S), hydroxyphenyl (H)
Bonding in subcomponent	Carbon-oxygen and carbon-carbon bond, mainly b-O-4 ether bonding
Polymerization	4000
Bonding between three components	Chemical bonding with hemicelluloses
Structure	3D, amorphous, non-homogeneous, nonlinear polymer

8.2.2 Types of Lignin

Chemical properties like composition, molecular weight, and molecular structure, as well as physical properties like solubility and hydrophilicity/hydrophobicity of lignin are greatly affected by various pretreatment methods of biomass. On the basis of these processes, lignin can be categorized as (Table 8.2):

i) Lignosulfonate Lignin
It is obtained from sulfite processes where alkaline earth metal sulfite mixture is used. In this type of lignin, sulfonate groups are mainly present at a position of the propyl side chain. It is soluble in water, making it different from other types of lignin. It has various applications like dispersing agent, surfactant, stabilizer, and adhesive and it is mainly because of the presence of high dense functional groups having unique colloidal properties. This type of lignin has relatively high molecular weight and contains a large quantity of ash with a considerable amount of carbohydrates. This lignin obtained from hard and softwoods is commercially available.

ii) Kraft Lignin
This type of lignin is condensed with strong ether bonding and is high in refractory CeC bonding such as biphenyl and methylenebridge. It has some impurities in its structure such as covalent bonding with sulfur species, mainly thiols. These impurities can further inhibit valorization due to the poisonous nature of sulfur which deactivates most of the catalysts. It has a large number of phenolic groups with a small amount of ash and carbohydrate. Kraft

Table 8.2 Types of lignin.

Type of lignin	Purity	Scale of production
Lignosulfonate	Low	Industrial
Kraft	Moderate	Industrial
Soda	Moderate-Low	Industrial
Organosolv	High	Pilot/Demo

lignin is also commercially available from hard and softwoods (24).

iii) Soda Lignin
Soda lignin is very much different from lignosulfonate and kraft lignin because of its sulfur-free nature and the presence of vinyl ethers. This lignin is made more refractory and is condensed due to harsh pulping conditions. It has minor traces of ash and carbohydrates. In contrast, it has an opposite process as compared to kraft lignin, where lignin is gained under acidic conditions. It is also commercially available but only from a small number of annual hardwood crops [24].

iv) Organosolv Lignin
Its structure mainly depends on the pretreatment conditions under which organosolv lignin is obtained. Its structure is more or less similar to the native lignin, as it also contains many easily breakable aryl ether linkages (b-O-4). It has low molecular weight and polydispersity and is homogeneous in structure. In addition, this can be chemically changed into soda or Kraft lignin. Softwoods and hardwoods are utilized for the production of lignin by using pilot, as well as demonstration plant scale, and it is also commercially available [23].

8.2.3 Applications of Lignin

Lignin has many useful properties, i.e., it is biodegradable, has antioxidant and antimicrobial activities, etc. Thus, it is used in different industries as an important source for liquid fuel production and commercially important chemicals and products. It is studied that the high density of lignin makes it an important source for production of biofuel from lignin feedstock. It is projected that around 50 million tons of lignin was produced by the paper and pulp industry in 2010, but commercially only 2% of it was used [25]. Lignin is one of the waste products obtained during bioethanol production, though some of its fraction is used for production of char. During biofuel production, this lignin feedstock gets properly utilized. Lignin is also used as an adsorbent agent, an emulsifier, a chelating agent, a dispersant reagent in cement, and as gypsum beads [25, 26]. Lignin is a source for the production of bioplastics as well as biofertilizers and also serves as energy storage devices such as high energy super capacitors [27].

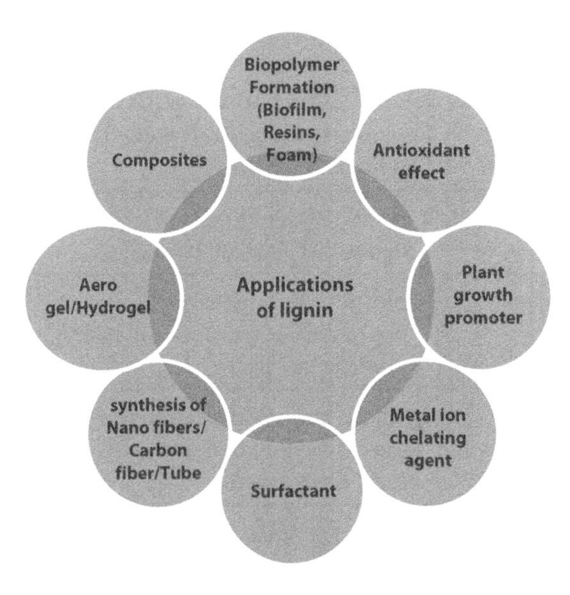

Figure 8.2 Applications of lignin.

The industrial applications of lignin are restricted due to their multifarious nature and undefined chemical structure. For example, kraft lignins which are commercially purchased from softwoods may have diverse compositions as well as structures [28]. In addition, the difficulty in analysis of structure and isolation of lignin from LCBM is increased by lignin-carbohydrate complex (LCC) [29]. The ultimate goal for the biorefineries are the value-added consumption of lignin and its degradation products, therefore its structural determination is necessary so that theoretical direction on constructing and optimizing degradation processes can be determined, generating valuable aromatic chemicals which act as low-molecular-mass feedstocks [30], determining the economic viability. Biopolymers like gelatin and alginate were blended with lignin to develop active biodegradable films for packaging of food stuffs as well as biomedical applications [31].

Lignin is able to scavenge radicals introduced by irradiation because of its aromatic structure and therefore acts in lignocellulosic materials as a UV stabilizer and an anti-oxidant. NovaFiber lignin may be used as a functional additive in plastics like polypropylene and polyethylene. They provide value addition to the plastics in two diverse ways:

- Lignin can act as filler to reduce the price of the compound and this is because of its low cost price; this way, it serves a

Table 8.3 Various uses of lignin.

Use	Comments	Reference
Fuel	Recovery boiler in Kraft processes Bioethanol production	[32] [33]
Animal feed additive	for production of feed stuffs, lignosulfonates can be used	[34]
Concrete additive	Lignosulfonates increase the pourability of concrete and have plasticizing effect	[35]
Syngas	Synthesis gas can be derived from the gasification of lignin recovered from Kraft process	[32]
Production of Vanillin	It can be produced by chemical processing of softwood lignin from pulping of sulfite	[36]
Adhesives and Resins	Lignin partly replaces phenols in different types of resins and adhesives	[37]

substitute to conventional fillers like talc, chalk, and carbon black

- Lignin can act as a high-end additive and this is because of its UV-stabilizing properties; this way it is a substitute to conventionally applied stabilizers like HALS (Hindered Amine Light Stabilizer). These stabilizers are chemicals which are expensive compared to lignin [38].

8.3 Lignin as Fuel

For the sustainable development of society, there is a need to employ inedible biomass for the production of valuable products through biorefinery operations (fuels, chemicals and materials). Thus, lignocellulose, which is composed of cellulose, hemicellulose, and lignin, is a potential feedstock for biorefineries to produce biofuels and biochemicals [39]. Lignocellulose is commonly accessible, carbon-neutral, and an uneatable bioresource that can be regenerated on a world-wide basis in huge quantities every year. However, nearly all the biorefinery processes at present focus on the utilization of carbohydrate fractions (cellulose and hemicellulose), leaving lignin, the second most abundant global polymer, underutilized. The development of 79 billion liters of second-generation biofuels annually by 2022 is consented by the U.S. Energy Security and Independence Act of 2007. 223 million tons of biomass will be used annually after consuming a yield of 355 L of bioethanol per dry ton of biomass, producing about 62 million tons of lignin. Therefore, about 60% more lignin is produced than what is needed to meet the internal energy use by its combustion [22, 33]. Thus, to make the industrial scale biorefinery plants cost-competitive, the efficient valorization of lignin into value-added chemicals or fuels is crucial [34, 35]. The only large-volume renewable aromatic feedstock is lignin, found in most terrestrial plants on average in a range of 15–30% by dry weight and 40% by energy [36, 37]. Traditionally, lignin has been burned for most of the large-scale industrial processes that use plant polysaccharides for the generation of the power needed for biomass conversion. However, because the wastes that are rich in lignin are wet and, as compared, lignin has lower energy content than coal, its value is limited under $50/dry ton.

Moreover, lignin is generated by most of the cellulosic ethanol plants in such amounts that it over addresses the energy requirement for this process. Thus, the overall biorefinery competitiveness is significantly enhanced by the conversion of lignin to higher value chemicals and fuels [42]. For example, reduction in the feedstock cost of the overall process can be achieved by the conversion of about 20% of the lignin in a plant containing around 25% lignin into products valued at about $1.00/lb, which would produce additional profit of about $100/ton of feedstock fed to a biological process.

Lignin has greater energy density than cellulose, as well as a higher carbon to oxygen ratio (above 2:1) which makes it a potential candidate for fuel and aromatics (e.g., phenol, benzene, toluene, and xylene) production [38, 39]. Although many studies have also focused on conversion of lignin to valuable products, only a few of these efforts can be commercially profitable because of their low quality of end product and lower yield products. In the global energy market, bio-energy has been a major component due to its high bio-energy productivity and low requirement for energy infrastructure modification. Biodiesel production from lipid-rich biomass, bio-ethanol production from sugar-rich biomass, and biogas production from volatile-matter-dense feedstock have been commercialized [40]. A considerable amount of biomass waste is still discarded due to its low-end product values and high processing costs. This includes plants, agrowastes, de-oiled seed cakes, depleted or used up grains, food wastes, forest wastes, digestated residuals, and municipal wastes. These types of wastes usually contain less digestible compounds (e.g., sugars and fatty acids) and refractory lignin, which requires more harsh reaction conditions for the extraction of valuable compounds [41]. Thus, pyrolysis is an inexpensive and potential extraction option for these compounds with easy handling and operation and high compatibility with diverse feedstocks [45, 46].

The most efficient form of biomass-derived energy is biofuel due to the economically competitive processes and the drop by character of the fuels relative to the most popular fossil-based fuels [47]. Standard biofuel mainly contains mono-alkyl esters of long chain fatty acids derived from energy crops, lignocellulose-derived components, animal fats and vegetable oils [48].

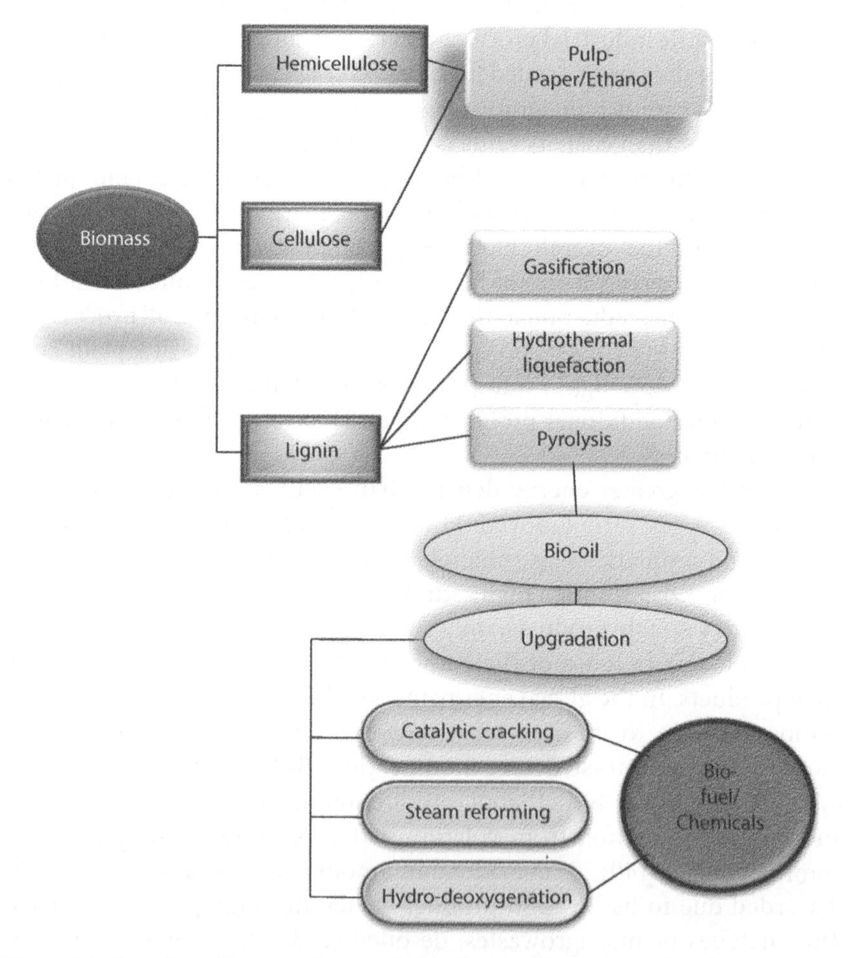

Figure 8.3 Overview of lignin utilization for bio-fuel or chemicals production.

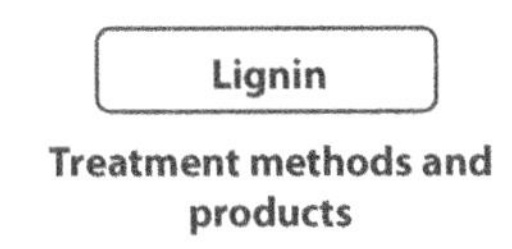

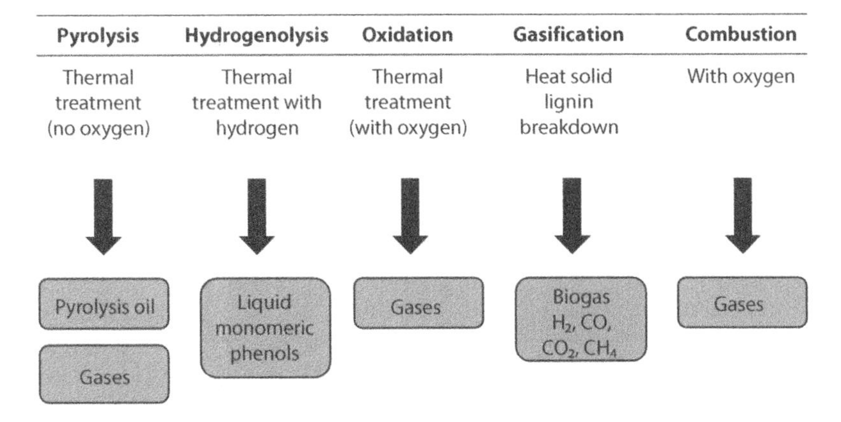

Figure 8.4 Depolymerization of lignin by chemical methods and its use in production of fuel.

8.3.1 Bioethanol Production

Biomass-derived biofuels, bioethanol, and specially high-value transportation fuels have received great attention nowadays [43, 44]. Thus, bioethanol production is considered to be a sustainable energy technology, which can lower the high demand of crude oil and also improves some environmental damages [50]. For bioethanol production, lignocellulosic biomass is found to be the most promising feedstock due to its abundance in the environment [51]. Smokeless burning and low flash point makes bioethanol an important biofuel [52]. Pre-treatment of biomass is done for bioethanol production and it is crucial because of the special structure of lignocelluloses which causes "recalcitrance" to enzyme and microbial deconstruction [48, 49]. For the production of bioethanol, second generation biorefineries use pre-treatment technologies to make the polycarbohydrates available for enzymatic hydrolysis [55]. Bioethanol has drawn attention to itself in this century as a renewable fuel and also serves as a platform compound for chemicals [53, 54]. As a second generation biomass for bioethanol production, lignocellulosic biomass is considered to be a promising candidate for production [51] because bioethanol produced by lignocelluloses does not compete with food production. The unused woods and biomass can be easily utilized for bioethanol production. Two methods are primarily used for production of bioethanol using lignocelluloses with the cellulolytic enzyme, cellulase. One of the methods used is separate hydrolysis and fermentation (SHF) to produce fermentable sugars followed by fermentation with yeast. Another method is simultaneous saccharification and fermentation (SSF). Many additives were also used to enhance the enzymatic saccharification efficiency, like surfactants which were developed since the 1980's [56]. In the 2010's, lignosulfonates, isolated lignin [57], extracted lignin, and organosolv lignin [58] were also studied for improving enzymatic saccharification of lignocelluloses without further modification [59]. There are four steps involved in the conversion of lignocelluloses to bioethanol: a physico-chemical pretreatment [60], saccharification [61], fermentation, and distillation [62]. Enzymatic hydrolysis is used more than acid hydrolysis for the saccharification of polysaccharides from lignocellulosic biomass because enzymatic hydrolysis can be carried out under milder conditions and specialized reaction vessels are not required [58, 59]. There is a disadvantage to using enzymatic hydrolysis, i.e., the high cost of cellulolytic enzyme cocktails, which is a combination of cellulases, hemicellulases, and other enzymes with different substrate activities [65].

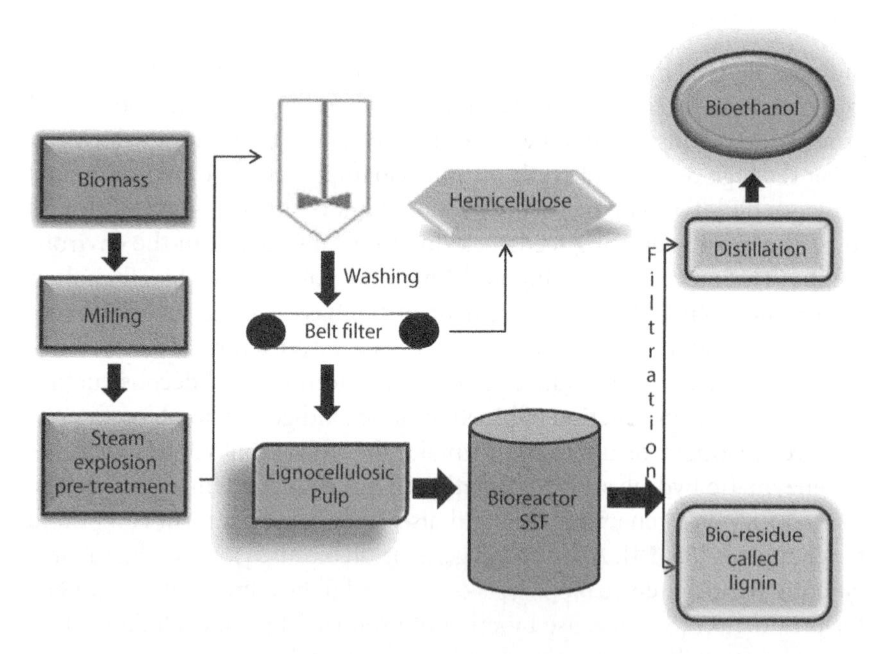

Figure 8.5 Processes used to produce bioethanol; Bio-residue called lignin is a co-product, as is hemicellulose.

It is thus required to maximize activity over a prolonged reaction time in multiple rounds of saccharification and also to improve the cellulolytic enzyme activities to accomplish cost effective enzymatic saccharification [66]. These days, maize plant (*Zea maysL.*), a lignocellulosic biomass, is considered as an agricultural waste as it is very difficult to access the polysaccharides stocked in the cell wall. This mainly hinders its interaction with lignin, as well as being necessary to improve the quality of maize lignocelluloses, so that a large amount of biomass can be used as a source of cellulosic bioethanol [67].

8.3.2 Bio-Oil Production

Pyrolysis of lignocellulosic biomass leads to the production of second era liquid fuel, generally known as bio-oil. The bio-oil produced is a dark brown, red, or green colored liquid depending on the chemical composition of biomass [63, 64]. As compared to the original biomass density, it has 5 to 20 times higher energy density. It is not suitable for direct use as transportation fuel because of different physical and chemical properties, however, it can be co-processed with the petroleum-derived feedstocks [24].

Physical Properties of Bio-Oil

Bio-oil is non-homogenous, which leads to fractional separation of phases and filtration issues. When contrasted with fossil fuel, it possesses poor fuel properties such as low pH, high density, instability, high corrosiveness, and low heating values [70]. Here, the reason for corrosion is the acidic character and low pH of bio-oil. The heating values and flame temperature are significantly reduced by the presence of high water content in bio-oil. It also causes a decrease in combustion rate and ignition delay [71]. Due to the deposition of alkali metals in bio-oil, this results in corrosion in the combustion area.

Chemical Properties of Bio-Oil

Bio-oil is a complex mixture, usually containing more than 100 organic compounds across a variety of functionalities, like carbohydrates, aldehydes, ketones, acids, pyrans, furans, aromatics, and hydrocarbons [66]. Bio-oil does not easily mix with the hydrocarbons and is less steady and the reason behind this is its polar nature. The decomposition of hemicellulose and cellulose leads to the structural formation of sugars, alcohols, acids, esters, furans, aldehydes, and ketones [67, 68]. The decomposition of lignin results in formation of phenolic compounds where phenols are substituted with aldehyde, alcohol, carboxylic acid, etc. [74]. The major phenolic compounds obtained from the pyrolysis of lignin are cresol, catechol, anisole, guaiacol, and syringol [66]. The contribution of phenolic compounds in bio-oil production is around 25-30%. In addition to this, they also have inferior oxygen content. So, the upgradation of phenol and phenolic derivatives has paid substantial attention to achieve the potential future fuel [24].

8.3.3 Syngas Production

Syngas is a mixture of hydrogen and carbon monoxide (H_2 + CO) which can be obtained by gasification of different materials. Several other compounds like ethanol, natural gas, and glycerol have also been found to produce syngas or hydrogen in several processes [69, 70]. Lignin-rich residues obtained from second generation bioethanol production are used for production of syngas that can be useful in the gas fermentation process [55]. The syngas obtained from gasification of lignin-rich biorefinery residues offers the potential to produce higher added-value products, such as liquid fuels and chemicals [77], but this has been assessed only to a limited extent on the bench and pilot scale [72–74]. A syngas cycle was proposed by Rostrup-Nielsen which involves steam reforming and gasification to produce syngas from complex materials [80, 81]. Gasification of lignocellulosic biomass is considerably different from gasification of lignin because lignin has different physicochemical characters. The chemical composition and structure of lignin, i.e., an aromatic polymer with higher C/O ratio than lignocellulosic biomass, favors formation of tar [82]. Supercritical water gasification is the most used method for thermochemical conversion of biomass into hydrogen-rich gas [75, 76]. In this method, at lower temperatures wet biomass are often converted into the hydrogen-rich syngas with higher efficiencies due to the high dispersion and effective heat transfer of water in its supercritical condition [83].

8.4 As Fuel Additive

For a long time, many fundamental researches have been done for conversion of lignin to important value added chemicals, alternative fuels, etc. However, only a few of them have been commercialized [23]. It has been a great challenge so far for its effective conversion and thus is burnt to recover energy in the paper and pulp industry [78, 79]. Various limiting factors of lignin utilization are its (i) hydrophobicity (the effective interaction and reactions between lignin macromolecules and reactants which are hindered by the non-covalent π–π interaction between the aromatic rings in lignin) [85, 86], (ii) indigestibility (the cross-linked molecular structure of lignin results in significant chemical resistance toward depolymerization) [87], and (iii) complex chemical structures (lignin structure and its property depend on the source by nature) [88].

The polymeric structure of lignin is broken down into low molecular-weight aromatic compounds, which is the first and crucial approach for lignin utilization [83, 84]. Lignin based additives were prepared by coupling different types of isolated lignins like organosolv lignins and kraft and soda lignins with epoxylated polyethylene glycol (PEG) [91]. The lignin conversion methods can be classified as acid/base catalyzed depolymerization/hydrolysis, pyrolysis, hydrotreating (hydrodeoxynation, hydrogenation, hydrogenolysis, and hybrid processes therein), chemical oxidation, liquid phase reforming, gasification, and biodegradation [92]. Hydrolysis reactions are either catalyzed by acid (typically at 0–200 °C) or base (100–300 °C) that breaks the C–O and/or C–C bonds in lignin and generate monomeric phenols [86, 87]. In the absence of oxygen, thermal pyrolysis is generally operated above 450°C, which produces bio-oil from lignin directly [88, 89], however in the presence of water (solvent), hydrothermal liquefaction is operated at high pressure and medium temperature, which produces bulk aromatic compounds [90–92]. At a lower temperature between 0–250°C, an oxidation reaction takes place and produces aldehydes, acids, alcohols, and platform chemicals [99]. Cyclohexanol which is produced by lignocellulosic biomass is used as a fuel additive and is mixed with diesel and synthetic diesel [96, 97]. It is also used in gas to liquid (GTL)-diesel blend. It is more effective in combination with a GTL-diesel blend and also improves particulate matter (PM) [93, 95]. In comparison to conventional diesel fuel, it also showed a reduction in NOx [98]. This blend has shown significant potential as a fuel and its properties fall within EN590's specifications for a diesel fuel [94, 100].

8.5 Conclusion

As the petroleum reserves are exhausted, the demand of fuel is also increasing gradually. So, to deal with this problem, agrowaste biomass can be considered as an appropriate alternative. Thus, lignin is the second most abundant polymer and is used as a source of many important chemicals and fuels. It also serves as a standalone alternative fuel. It is an irregular three-dimensional, amorphous, and highly branched phenolic polymer and accounts for 15–30% by weight of lignocellulosic biomass. Lignin has diverse applications like biopolymer formation (biofilm, resins, and foam), antioxidant effect, composites, an adsorbent agent, as an emulsifier, as a chelating agent, as a dispersant reagent in cement, etc. Lignin separation is based on different biomass types and compositions where its amount varies from biomass to biomass and pretreatment methods (biological, mechanical, chemical, and physiochemical). In contrast, lignin is a waste material of the bioethanol and paper industries and is used as a precursor for the production of green fuel and chemicals. There are some standard methods for bio-oil production from lignin, i.e., pyrolysis and liquefaction. Lignin based bio-oil is unstable, corrosive, non-homogenous, and mainly contains phenol. It also contains carbon and hydrogen which is analogous to fossil fuels and has great possibility for production of fuel. Therefore, lignin is used for production of fuels and fuel additives.

References

1. Agwu, O.E., Akpabio, J.U., Using agro-waste materials as possible filter loss control agents in drilling muds: A review. *Journal of Petroleum Science and Engineering*, doi: 10.1016/j.petrol.2018.01.009, 2018.
2. Chandra, R., Takeuchi, H., Hasegawa, T., Hydrothermal pretreatment of rice straw biomass: a potential and promising method for enhanced methane production. *Appl Energy.*, 94, 129–40, 2012.
3. Kauldhar, B.S., Yadav, S.K., Turning waste to wealth: A direct process for recovery of nano-silica and lignin from paddy straw agro-waste. *J Clean Prod.*, 194, 158–66, 2018.
4. Gazliya, N., Aparna, K., Microwave-assisted alkaline delignification of banana peduncle. *J Nat Fibers.*, 1–10, 2019.
5. Nazimudheen, G., Sekhar, N.C., Sunny, A., Kallingal, A., Hasanath, B., Physiochemical characterization and thermal kinetics of lignin recovered from sustainable agrowaste for bioenergy applications. *Int J Hydrogen Energy.* doi.org/10.1016/j.ijhydene.2020.03.172, 2020.
6. Barakate, A., Stephens, J., Goldie, A., Hunter, W.N., Marshall, D., Hancock, R.D., Syringyl lignin is unaltered by severe sinapyl alcohol dehydrogenase suppression in tobacco. *Plant Cell.*, 23(12), 4492–506, 2011.
7. Thielemans, W., Wool, R.P., Butyrated kraft lignin as compatibilizing agent for natural fiber reinforced thermoset composites. *Compos part A Appl Sci Manuf.*, 35(3), 327–38, 2004.
8. Boerjan, W., Ralph, J., Baucher, M., Lignin biosynthesis. *Annu Rev Plant Biol.*, 54(1), 519–46, 2003.
9. Vanholme, R., Morreel, K., Ralph, J., Boerjan, W., Lignin engineering. *Curr. Opin. Plant Biol.*, 278–85, 2008.

10. Lourenço, A., Rencoret, J., Chemetova, C., Gominho, J., Gutiérrez, A., Del Río, J.C., Lignin composition and structure differs between Xylem, Phloem and phellem in quercus suber L., *Front Plant Sci.*, 7, 1612, 2016.
11. Lu, Y., Lu, Y.C., Hu, H.Q., Xie, F.J., Wei, X.Y., Fan, X., Structural characterization of lignin and its degradation products with spectroscopic methods. *J Spectrosc.*, 2017, 15, 2017.
12. Wang, H., Pu, Y., Ragauskas, A., Yang, B., Bioresource Technology From lignin to valuable products – strategies , challenges , and prospects. *Bioresour Technol.*, 0-1, 2018.
13. Aadil, K.R., Jha, H., Physico-chemical properties of lignin–alginate based films in the presence of different plasticizers. *Iran Polym J.*, 25(8), 661–70, 2016.
14. Ralph, J., Lundquist, K., Brunow, G., Lu, F., Kim, H., Schatz, P.F., Lignins: natura Microwave-assisted alkaline delignification of banana peduncle l polymers from oxidative coupling of 4-hydroxyphenyl-propanoids. *Phytochem Rev.*, 3(1), 29–60, 2004.
15. Grabber, J.H., Schatz, P.F., Kim, H., Lu, F., Ralph, J., Identifying new lignin bioengineering targets: 1. Monolignol-substitute impacts on lignin formation and cell wall fermentability. *BMC Plant Biol.*, 10(1), 1–13, 2010.
16. Boudet, A.M., Lignins and lignification: selected issues. *Plant Physiol Biochem.*, 38(1–2), 81–96, 2000.
17. Buranov, A.U., Mazza, G., Lignin in straw of herbaceous crops. *Ind Crops Prod.*, 28(3), 237–59, 2008.
18. Kai, D., Tan, M.J., Chee, P.L., Chua, Y.K., Yap, Y.L., Loh, X.J., Towards lignin-based functional materials in a sustainable world. *Green Chem.*, 18(5), 1175–2000, 2016.
19. Zeng, Y., Himmel, M.E., Ding, S.Y., Coherent Raman microscopy analysis of plant cell walls. *Biomass conversion.*, 908, 49–60, 2012.
20. Zhou, C., Li, Q., Chiang, V.L., Lucia, L.A., Griffis, D.P., Chemical and spatial differentiation of syringyl and guaiacyl lignins in poplar wood via time-of-flight secondary ion mass spectrometry. *Anal Chem.*, 83(18), 7020–6, 2011.
21. Jung, S., Foston, M., Kalluri, U.C., Tuskan, G.A., Ragauskas, A.J., 3D chemical image using TOF-SIMS revealing the biopolymer component spatial and lateral distributions in biomass. *Angew Chemie.*, 124(48), 12171–4, 2012.
22. Upton, B.M., Kasko, A.M., Strategies for the conversion of lignin to high-value polymeric materials: review and perspective. *Chem Rev.*, 116(4), 2275–306, 2016.
23. Ragauskas, A.J., Beckham, G.T., Biddy, M.J., Chandra, R., Chen, F., Davis, M.F., Lignin valorization: improving lignin processing in the biorefinery. *Science*, 344, 6185, 2014.
24. Kumar, A., Kumar, J., Bhaskar, T., Utilization of lignin : A sustainable and eco-friendly approach. *J Energy Inst.*, https://doi.org/10.1016/j.joei.2019.03.005, 2019.

25. Smolarski, N., High-value opportunities for lignin: unlocking its potential. *Frost & Sullivan.*, 1, 2012.
26. Yang, H., Yan, R., Chen, H., Lee, D.H., Zheng, C., Characteristics of hemicellulose, cellulose and lignin pyrolysis. *Fuel.*, 86(12–13), 1781–8, 2007.
27. Sena, M. G., Almeida,V. E., Duarte, J.C., Eco-friendly new products from enzymatically modified industrial lignins. *Ind Crops Prod.*, 27(2), 189–95, 2008.
28. Dodd, A.P., Kadla, J.F., Straus, S.K., Characterization of fractions obtained from two industrial softwood kraft lignins. *ACS Sustain Chem Eng.*, 3(1),103–10, 2015.
29. Giummarella, N., Zhang, L., Henriksson, G., Lawoko, M., Structural features of mildly fractionated lignin carbohydrate complexes (LCC) from spruce. *Rsc Adv.*, 6(48), 42120–3, 2016.
30. Wang, H., Ben, H., Ruan, H., Zhang, L., Pu, Y., Feng, M., Effects of lignin structure on hydrodeoxygenation reactivity of pine wood lignin to valuable chemicals. *ACS Sustain Chem Eng.*, 5(2), 1824–30, 2017.
31. Aadil, K.R., Pandey, N., Mussatto, S.I., Jha, H., Green synthesis of silver nanoparticles using acacia lignin, their cytotoxicity, catalytic, metal ion sensing capability and antibacterial activity. *J Environ Chem Eng.*, 7(5), 103296, 2019.
32. Naqvi, M., Yan, J., Dahlquist, E., Bio-refinery system in a pulp mill for methanol production with comparison of pressurized black liquor gasification and dry gasification using direct causticization. *Appl Energy.*,90(1), 24–31, 2012.
33. Zhang, J., Liu, J., Kou, L., Zhang, X., Tan, T., Bioethanol production from cellulose obtained from the catalytic hydro-deoxygenation (lignin-first refined to aviation fuel) of apple wood. *Fuel.*, 250, 245–53, 2019.
34. Doherty, W.O.S., Mousavioun, P., Fellows, C.M., Value-adding to cellulosic ethanol: Lignin polymers. *Ind Crops Prod.*,33(2), 259–76, 2011.
35. Plank, J., Applications of biopolymers and other biotechnological products in building materials. *Appl Microbiol Biotechnol.*,66(1), 1–9, 2004.
36. Pacek, A.W., Ding, P., Garrett, M., Sheldrake, G., Nienow, A.W., Catalytic conversion of sodium lignosulfonate to vanillin: engineering aspects. Part 1. Effects of processing conditions on vanillin yield and selectivity. *Ind Eng Chem Res.*,52(25), 8361–72, 2013.
37. Stewart, D., Lignin as a base material for materials applications: Chemistry, application and economics. *Ind Crops Prod.*,27(2), 02–7, 2008.
38. Gosselink, R.J.A., Snijder, M.H.B., Kranenbarg, A., Keijsers, E.R.P., Jong, E. D., Stigsson, L.L., Characterisation and application of NovaFiber lignin. *Industrial Crops and Products.*, 20, 191–203, 2004.
39. Rinaldi, R., Jastrzebski, R., Clough, M.T., Ralph, J., Kennema, M., Bruijnincx, P.C.A., Paving the way for lignin valorisation: recent advances in bioengineering, biorefining and catalysis. *Angew Chemie Int Ed.*, 55(29), 8164–215, 2016.

40. Shen, D., Jin, W., Hu, J., Xiao, R., Luo, K., An overview on fast pyrolysis of the main constituents in lignocellulosic biomass to valued-added chemicals : Structures, pathways and interactions. *Renew Sustain Energy Rev.*, 51, 761–74, 2015.
41. Xu, C., Arancon, R.A.D., Labidi, J., Luque, R., Lignin depolymerisation strategies: towards valuable chemicals and fuels. *Chem Soc Rev.,* 43(22), 7485–500, 2014.
42. Gillet, S., Aguedo, M., Petitjean, L., Morais, A.R.C., da Costa Lopes AM, Łukasik, R.M., Lignin transformations for high value applications: towards targeted modifications using green chemistry. *Green Chem.*, 19(18), 4200–33, 2017.
43. Lupoi. J.S., Singh, S., Parthasarathi, R., Simmons, B.A., Henry, R.J., Recent innovations in analytical methods for the qualitative and quantitative assessment of lignin. *Renew Sustain Energy Rev.*, 49, 871–906, 2015.
44. Laskar, D.D., Yang, B., Wang, H., Lee, J., Pathways for biomass-derived lignin to hydrocarbon fuels. *Biofuels, Bioprod Biorefining.*, 7(5), 602–26, 2013.
45. Schutyser, W., Renders, T., Van, D. B. S., Koelewijn, S. F., Beckham, G.T., Sels, B.F., Chemicals from lignin: an interplay of lignocellulose fractionation, depolymerisation, and upgrading. *Chem Soc Rev.*, 47(3), 852–908, 2018.
46. Cheng, F., Bayat, H., Jena, U., Brewer, C.E., Impact of feedstock composition on pyrolysis of low-cost, protein- and lignin-rich biomass: A review. *J Anal Appl Pyrolysis*, 147, 104780, 2020, https://doi.org/10.1016/j.jaap.2020.104780.
47. Bhuiya, M.M.K., Rasul, M.G., Khan, M.M.K., Ashwath, N., Azad, A.K., Hazrat, M.A., Prospects of 2nd generation biodiesel as a sustainable fuel–Part 2: Properties, performance and emission characteristics. *Renew Sustain Energy Rev.*, 55, 1129–46, 2016.
48. Janaun, J., Ellis, N., Renewable and Sust. *Energy Rev.*, 14, 1312, 2010.
49. Yoo, C.G., Dumitrache, A., Muchero, W., Natzke, J., Akinosho, H., Li, M., Significance of Lignin S/G Ratio in Biomass Recalcitrance of Populus trichocarpa Variants for Bioethanol Production. *ACS Sustain Chem Eng.*, 6(2), 2162–8, 2018.
50. Sarkar, N., Ghosh, S.K., Bannerjee, S., Aikat, K., Bioethanol production from agricultural wastes: an overview. *Renew energy.*, 37(1), 19–27, 2012.
51. Olofsson, K., Bertilsson, M., Lidén, G., A short review on SSF–an interesting process option for ethanol production from lignocellulosic feedstocks. *Biotechnol Biofuels.*, 1(1), 1–14, 2008.
52. Baeyens, J., Kang, Q., Appels, L., Dewil, R., Lv, Y., Tan, T., Challenges and opportunities in improving the production of bio-ethanol. *Prog Energy Combust Sci.,* 47, 60–88, 2015.
53. Zhu, J.Y., Pan, X.J., Woody biomass pretreatment for cellulosic ethanol production: technology and energy consumption evaluation. *Bioresour Technol.*, 101(13), 4992–5002, 2010.

54. Zhu, J.Y., Pan, X., Zalesny, R.S., Pretreatment of woody biomass for biofuel production: energy efficiency, technologies, and recalcitrance. *Appl Microbiol Biotechnol.*, 87(3), 847–57, 2010.
55. Liakakou, E.T., Vreugdenhil, B.J., Cerone, N., Zimbardi, F., Pinto, F., André, R., Gasification of lignin-rich residues for the production of biofuels via syngas fermentation: Comparison of gasification technologies. *Fuel.*, 251, 580–92, 2019.
56. Seo, D. J., Fujita, H., Sakoda, A., Effects of a non-ionic surfactant, Tween 20, on adsorption/desorption of saccharification enzymes onto/from lignocelluloses and saccharification rate. *Adsorption.*, 17(5), 813–22, 2011.
57. Lou, H., Zhou, H., Li, X., Wang, M., Zhu, J.Y., Qiu, X., Understanding the effects of lignosulfonate on enzymatic saccharification of pure cellulose. *Cellulose.*, 21(3), 1351–9, 2014.
58. Lai, C., Tu, M., Li, M., Yu, S., Remarkable solvent and extractable lignin effects on enzymatic digestibility of organosolv pretreated hardwood. *Bioresour Technol.*, 156, 92–9, 2014.
59. Cheng, N., Yamamoto, Y., Koda, K., Tamai, Y., Uraki, Y., Amphipathic lignin derivatives to accelerate simultaneous saccharification and fermentation of unbleached softwood pulp for bioethanol production. *Bioresour Technol.*, 173, 104–9, 2014.
60. Uppugundla, N., da Costa Sousa L, Chundawat,S.P.S., Yu, X., Simmons, B., Singh, S., A comparative study of ethanol production using dilute acid, ionic liquid and AFEX™pretreated corn stover. *Biotechnol Biofuels.*, 7(1), 1–14, 2014.
61. Kumar, D., Murthy, G.S., Stochastic molecular model of enzymatic hydrolysis of cellulose for ethanol production. *Biotechnol Biofuels.*, 6(1), 1–20, 2013.
62. Phillips, R.B., Jameel, H., Chang, H.M., Integration of pulp and paper technology with bioethanol production. *Biotechnol Biofuels.*, 6(1), 1–12, 2013.
63. Sun, Y., Cheng, J., Hydrolysis of lignocellulosic materials for ethanol production: a review. *Bioresour Technol.*, 83(1), 1–11, 2002.
64. Cheng, F., Brewer, C.E., Producing jet fuel from biomass lignin: Potential pathways to alkyl-benzenes and cycloalkanes. *Renew Sustain Energy Rev.*, 72, 673–722, 2017.
65. Nguyen, Q.A., Saddler, J.N., An integrated model for the technical and economic evaluation of an enzymatic biomass conversion process. *Bioresour Technol.*, 35(3), 275–82, 1991.
66. Deshpande, M. V., Eriksson, K.E., Reutilization of enzymes for saccharification of lignocellulosic materials. *Enzyme Microb Technol.*, 6(8), 338–40, 1984.
67. Fornalé, S., Capellades, M., Encina, A., Wang, K., Irar, S., Lapierre, C., Altered lignin biosynthesis improves cellulosic bioethanol production in transgenic maize plants down-regulated for cinnamyl alcohol dehydrogenase. *Mol Plant.*, 5(4), 817–30, 2012.

68. Gollakota, A.R.K., Reddy, M., Subramanyam, M.D., Kishore, N., A review on the upgradation techniques of pyrolysis oil. *Renew Sustain Energy Rev.*, 58, 1543–68, 2016.
69. Liu, W., You, W., Sun, W., Yang, W., Korde, A., Gong, Y., Ambient-pressure and low-temperature upgrading of lignin bio-oil to hydrocarbons using a hydrogen buffer catalytic system. *Nat Energy*, 5(10), 759–67, 2020.
70. Lee, H., Kim, H., Yu, M.J., Ko, C.H., Jeon, J.K., Jae, J., Catalytic hydrodeoxygenation of bio-oil model compounds over Pt/HY catalyst. *Sci Rep.*, 6(1), 1–8, 2016.
71. Saidi, M., Samimi, F., Karimipourfard, D., Nimmanwudipong, T., Gates, B.C., Rahimpour, M.R., Upgrading of lignin-derived bio-oils by catalytic hydrodeoxygenation. *Energy Environ Sci.*, 7(1), 103–29, 2014.
72. Dickerson, T., Soria, J., Catalytic fast pyrolysis: a review. *Energies.*, 6(1), 514–38, 2013.
73. Nanda, S., Mohanty, P., Kozinski, J.A., Dalai. A.K., Physico-chemical properties of bio-oils from pyrolysis of lignocellulosic biomass with high and slow heating rate. *Energy Env Res.*, 4(3), 21, 2014.
74. Huber, G.W., Iborra, S., Corma, A., Synthesis of transportation fuels from biomass: chemistry, catalysts, and engineering. *Chem Rev.*,106(9), 4044–98, 2006.
75. Wang, H., Wang, X., Li, M., Li, S., Wang, S., Ma, X., Thermodynamic analysis of hydrogen production from glycerol autothermal reforming. *Int J Hydrogen Energy.*, 34(14), 5683–90, 2009.
76. Rossi, C., Alonso, C.G., Antunes, O.A.C., Guirardello, R., Cardozo, F. L., Thermodynamic analysis of steam reforming of ethanol and glycerine for hydrogen production. *Int J Hydrogen Energy.*, 34(1), 323–32, 2009.
77. Pandey, M.P., Kim, C.S., Lignin depolymerization and conversion: a review of thermochemical methods. *Chem Eng Technol.*, 34(1), 29–41. 2011.
78. Cerone, N., Zimbardi, F., Contuzzi, L., Alvino, E., Carnevale, M.O., Valerio, V., Updraft gasification at pilot scale of hydrolytic lignin residue. *Energy & fuels.*, 28(6), 3948–56, 2014.
79. Cerone, N., Zimbardi, F., Contuzzi, L., Prestipino, M., Carnevale, M.O., Valerio, V., Air-steam and oxy-steam gasification of hydrolytic residues from biorefinery. *Fuel Process Technol.*, 167, 451–61, 2017.
80. Pinto, F., André, R.N., Carolino, C., Miranda, M., Abelha, P., Direito, D., Effects of experimental conditions and of addition of natural minerals on syngas production from lignin by oxy-gasification: Comparison of bench- and pilot scale gasification. *Fuel.*, 140, 62–72, 2015.
81. Rodríguez, O.N.E., Mendoza,C. E.A., Castro M. A.J., Saucedo, L. J., Maya,Y. R., Rutiaga, Q. J.G., Simulation of syngas production from lignin using guaiacol as a model compound. *Energies.*, 8(7), 6705–14, 2015.
82. Maniatis, K., Beenackers, A., Tar protocols. IEA Bioenergy Gasification Task. *Biomass and Bioenergy.*, 18(1), 1–4, 2000.

83. Safari, F., Tavasoli, A., Ataei, A., Choi, J.K., Hydrogen and syngas production from gasification of lignocellulosic biomass in supercritical water media. *Int J Recycl Org Waste Agric.*, 4(2), 121–5, 2015.
84. Rahimi, A., Ulbrich, A., Coon, J.J., Stahl, S.S., Formic-acid-induced depolymerization of oxidized lignin to aromatics. *Nature.*, 515(7526), 249–52, 2014.
85. Mu, L., Shi, Y., Chen, L., Ji, T., Yuan, R., Wang, H., [N-Methyl-2-pyrrolidone] [C1–C4 carboxylic acid]: a novel solvent system with exceptional lignin solubility. *Chem Commun.*, 51(70), 13554–7, 2015.
86. Zakzeski, J., Bruijnincx, P.C.A., Jongerius, A.L., Weckhuysen, B.M., The catalytic valorization of lignin for the production of renewable chemicals. *Chem Rev.*, 110(6), 3552–99, 2010.
87. Hossain, M.M., Aldous, L., Ionic liquids for lignin processing: Dissolution, isolation, and conversion. *Aust J Chem.*, 65(11), 1465–77, 2012.
88. Watkins, D., Nuruddin, M., Hosur, M., Tcherbi, N.A., Jeelani, S., Extraction and characterization of lignin from different biomass resources. *J Mater Res Technol.*, 4(1), 26–32, 2015.
89. Fache, M., Boutevin, B., Caillol, S., Vanillin production from lignin and its use as a renewable chemical. *ACS Sustain Chem Eng.*, 4(1), 35–46, 2016.
90. Fache, M., Boutevin, B., Caillol, S., Epoxy thermosets from model mixtures of the lignin-to-vanillin process. *Green Chem.*, 18(3), 712–25, 2016.
91. Aso, T., Koda, K., Kubo, S., Yamada, T., Nakajima, I., Uraki, Y., Preparation of novel lignin-based cement dispersants from isolated lignins. *J Wood Chem Technol.*, 33(4), 286–98, 2013.
92. Li, C., Zhao, X., Wang, A., Huber, G.W., Zhang, T., Catalytic transformation of lignin for the production of chemicals and fuels. *Chem Rev.*, 115(21), 11559–624, 2015.
93. Riaz, A., Kim, C.S., Kim, Y., Kim, J., High-yield and high-calorific bio-oil production from concentrated sulfuric acid hydrolysis lignin in supercritical ethanol. *Fuel.*, 172, 238–47, 2016.
94. Liu, W.J., Jiang, H., Yu, H.Q., Thermochemical conversion of lignin to functional materials: a review and future directions. *Green Chem.*, 17(11), 4888–907, 2015.
95. Nsimba, R.Y., Mullen, C.A., West, N.M., Boateng, A.A., Structure–property characteristics of pyrolytic lignins derived from fast pyrolysis of a lignin rich biomass extract. *ACS Sustain Chem Eng.*, 1(2), 260–7, 2013.
96. Kang, S., Li, X., Fan, J., Chang, J., Hydrothermal conversion of lignin: a review. *Renew Sustain Energy Rev.*, 27, 546–58, 2013.
97. Lee, H., Jae, J., Ha, J.M., Suh, D.J., Hydro-and solvothermolysis of kraft lignin for maximizing production of monomeric aromatic chemicals. *Bioresour Technol.*, 203, 142–9, 2016.
98. Kloekhorst, A., Heeres, H.J., Catalytic hydrotreatment of alcell lignin using supported Ru, Pd, and Cu catalysts. *ACS Sustain Chem Eng.*, 3(9), 1905–14, 2015.

99. Hanson, S.K., Baker, R.T., Knocking on wood: base metal complexes as catalysts for selective oxidation of lignin models and extracts. *Acc Chem Res.*, 48(7), 2037–48, 2015.
100. Herreros, J.M., Jones, A., Sukjit, E., Tsolakis, A., Blending lignin-derived oxygenate in enhanced multi-component diesel fuel for improved emissions. *Appl Energy.*, 116, 58–65, 2014.

9

Fly Ash Derived Catalyst for Biodiesel Production

Trinath Biswal[1], Krushna Prasad Shadangi[2*] and Prakash Kumar Sarangi[3]

[1]*Department of Chemistry, Veer Surendra Sai University of Technology, Burla, Odisha, India*
[2]*Department of Chemical Engineering, Veer Surendra Sai University of Technology, Burla, Odisha, India*
[3]*College of Agriculture, Central Agricultural University, Imphal, Manipur, India*

Abstract

Fly ash produced due to burning of coal in thermal power plants for generation of energy is a significant industrial solid waste material. There are many research works carried out for eco-utilization of fly ash in various ways. This study is focused on originating, chemical composition, chemical properties, and catalytic activity of coal fly ash for production of biodiesel. The raw fly ash (RFA) after recycling can be used as a suitable solid catalyst because of various benefits as compared with other homogenous catalysts. The presence of aluminosilicate and some rare earth metals in coal burning fly ash considered it as a suitable candidate material for various catalytic, adsorption, and other extraction processes including fuel production. However, the existence of hazardous materials is a significant environmental problem, which must be resolved in order to avoid water, air, and soil contamination. Only 1-2% conversion of biodiesel was achieved from soybean oil by using RFA. The acid-treated fly ash (ATFA) was mixed with a metal hydroxide like KOH or NaOH and $Ca(OH)_2$ and was then calcined at a temperature of 700°C for about 3 hours to formulate the solid catalyst. This solid catalyst synthesized by the combination of NaOH with ATFA is labelled as SC-Na, which exhibits better performance compared to those obtained only by simple mixing ATFA with $Ca(OH)_2$ or NaOH and the optimum mass ratio of NaOH with ATFA was 3:1, which achieves 97.8% of biodiesel conversion. Most of the applied modified processes are hydrothermal, thermal treatment, and alkali activation, which enhances the structural, textural, and morphological properties. The form of active catalyst could

**Corresponding author*: krushnanit@gmail.com; kpshadangi_chemical@vssut.ac.in

Prakash Kumar Sarangi (ed.) Biorefinery Production of Fuels and Platform Chemicals, (203–232) © 2023 Scrivener Publishing LLC

be achieved by either ion exchange or impregnation technology. There are many catalytic systems based on fly ash used for production of biodiesel, but among these, the fly ash/CaO and zeolite derived fly ash method is more sustainable and beneficial.

***Keywords*:** Fly ash, heterogeneous catalyst, coal fly ash, biodiesel, RFA

9.1 Introduction

The dependence on fossil fuels is a great issue nowadays. The vast utilization of fossil fuels indicates that within a few years the resources of the fossil fuels (coal and crude petroleum oil) will be finished. The use of fossil fuels is not only a threat to our valuable stock resources, but also a great problem for our future generation from both an energy and environmental point of view. Hence, finding alternative renewable sources becomes a hot topic for worldwide researchers. It is well known that biodiesel is one of the alternative fuels which can be blended with commercial diesel to run the engine. Biodiesel has numerous benefits such as biodegradability, production of less % CO_2, less toxic hydrocarbons, lower SO_2 emission during burning, excellent lubrication properties, higher flash point, and high octane number. Therefore, proper modification in the process of production can be a proper competitor of diesel fuel. The utilization of biodiesel with diesel at proper composition does not need any development in the currently used diesel engine [1, 2]. With the use of advanced technology and modern science, biodiesel can be synthesized by adopting the following concepts:

- Heterogeneous or homogeneous transesterification catalyzed by base
- Transesterification and esterification catalyzed by acid
- Synthesis of biodiesel catalyzed by heterogeneous bifunctional solid catalysts
- Transesterification catalyzed by enzymes
- Supercritical methanolysis and deoxygenation [3, 4].

However, all these existing processes for production of biodiesel do not show satisfactory performance. Microwave and ultrasound assisted biodiesel production techniques are some of the processes that can be used to get a better yield and fuel properties for fuel oil. However, for the commercialization of the production of biodiesel, research is focussed on the development of new and improved catalytic systems, using a various feedstock of triacylglycerol (TAG), along with various modified continuous and batch reactor systems [5].

On the basis of cost and environmental sustainability, researchers are searching for synthesizing efficient catalytic systems from various kinds of waste materials. Several homogeneous and heterogeneous catalysts are being used to enhance the yield and properties of biodiesel. The use of homogenous acid and base catalysts can replace the use of heterogeneous catalysts such as oxides of alkaline and alkali earth metals, mixed oxides, zeolites, sulfonated solids, modified layer of double hydroxides, ion exchange resins, supported heteropolyacids, etc. [6] as it has several drawbacks such as soap formation, high corrosion, and less catalyst recovery.

For designing a new modern catalytic system to save our environment from drastic pollution load, scientists and researchers are focussing on various kinds of waste materials in which it was found that the fly ash ejected from coal fired power plants, animal and agricultural waste, fly ash obtained from biomass, and industrial solid wastes having high percentage of Ca (slug and mud) are exhibiting the potential source of catalytic properties [7]. These waste materials are highly toxic and a great negative impact to the environment. The use of such kinds of wastes is beneficial in two ways: the financial problem of disposal is solved and it is utilized as a catalyst to enhance biofuel production. Now, it was identified that fly ash from power plants on valorisation can be effectively used as a catalytic system for production of biodiesel. Zeolites are considered aluminosilicate materials, whose presence in the fly ash exhibits the catalytic property in the production of biodiesel [8, 9].

9.2 Coal Fly Ash: Resources and Utilization

Fly ash produced from coal or coal fly ash (CFA) is an important solid waste obtained due to burning of pulverized coal. It was found that about 70% of the residues obtained due to burning of coal are fly ash, boiler slug, bottom ash, and desulfurization residues of solid flue gas. The CFA is captured by an electrostatic precipitator (ESP), whereas the bottom ash is found surrounding hoppers under the air preheaters and economizers of huge pulverized coal boilers.

The particulates of CFA are the maximum produced lightweight particulate matter (PM) and its size ranges from 0.5 to 300 μm and has predominantly spherical shape particles. The various parameters that influence the properties of fly ash include chemical composition of the original coal feed for burning, mineral phases, condition of burning such as boiler temperature, size of feed coal, configuration of boiler temperature, and particulate control equipment. CFAs are normally a complex mixture of organic and inorganic compounds along with individual minerals or class of minerals. This constitutes the characterization and identification of the CFA [10, 11].

CFAs are normally a complex form of mixture having exclusive characteristics such as heterogeneity, multicomponents, and different chemical compositions of their organic, inorganic, and the fluid constituents. The major inorganic components almost 90–99% present in the CFAs are normally both crystalline and amorphous states of SiO_2, Al_2O_3, CaO, and Fe_2O_3 along with coal layers or coal-bearing rock layers. The CFAs also consist of different quantities of some specific rare earth elements such Ce, Gd, La, Nd, and Sm and some trace elements, mainly As, Se, Cd, and Cr, which are generally obtained from the original coal sample, which convert its significantly harmful components [12].

During the burning of pulverised coal, the C, N, and S present in the coal are converted into hazardous gaseous components such as CO, NO_x, and SO_x along with water vapour. The degradation of pyrite also produces SO_x and the degradation of carbonate minerals produces CO_2. The decomposition of the minerals generates the residual solid masses, which are collected in the form of bottom ash and fly ash. The influence of heat on key minerals of coal is as follows:

- Clay: Mullite ($3AI_20_3$+$2Si0_2$)
- Pyrite: Fe_2O_3 + SO_x
- Calcite: CaO + CO_2
- Siderite: FeO+ CO_2
- Quartz: Softens [13, 14].

On the basis of the formation and breaking of particles of different sizes, fusion, agglomeration, and composition indicates the final status of fly ash and bottom ash. The oxides FeO and CaO are comparatively not stable and possibly in reaction produce glassy states of aluminosilicates. The Ca present in the coal may be converted into fumes containing CaO or undergo chemical reactions with quartz and clays, leading to the formation of fly ash or small particles of slag.

Fly ash always contains some percentage of carbon in it, which depends upon the efficiency of grinding and the rate of combustion. Hence, fly ash is considered as a heterogeneous mixture and the main features of the particles present in it are hollow cenospheres. The agglomeration of tiny particles is observed by adhering it in larger size particles of various compositions.

The CFAs are normally divided into two classes, such as Class F and Class C, on the basis of their composition and from the coal in which it was formed. According to the American Society for Testing and Materials (ASTMs), 70% of the CFA of Class F is the combination of SiO_2, Al_2O_3, and Fe_2O_3, as compared to the 50% of Class C CFA. The class F category of the fly ash are considered as the actual pozzolanic material showing cementitious properties. The CFAs of Class C category is obtained from the sub-bituminous coals and lignite having a percentage of CaO in which more than 20% exhibits self-cementitious properties [15, 16].

The unique properties of pozzolanic Class F and cementitious Class C fly ash made it a suitable material for use as a binding agent to form clinker, substituting cement in the manufacturing of concrete. ASTM defines CFA fly ash as a fine powder comprising of typically glassy, spherical particulate matter, which is obtained due to the combustion of pulverized coal, without or with co-combustion substances having the pozzolanic properties and mainly consists of Al_2O_3 and SiO_2 [17].

Later on, a new classification system was devised on the basis of the concentration of ash forming elements on the basis of three composition criteria:

- On the basis of the sum of the existence of the oxides of Si, K, Al, and Ti
- On the basis of oxides of sum of Ca, S, Mg, and Na
- Fe_2O_3.

According to the concentration of the various constituents, coal fired fly ash is categorized into the following four different categories:

(i) Sialic
(ii) Calsialic
(iii) Ferricalsialic
(iv) Ferrisialic [18].

With the plentiful availability and comparability of low cost, coal is now treated as a potential global energy source and is the cause of generation of huge quantities of CFAs. The increase in the demand of energy extracted from coal is now continually increasing every year, for example, in 2005 the fly ash produced in the world was about 500 million tons, whereas in 2015 it is increased to 750 million tons. Among the total production of CFA, only 25% is used for various useful, eco-friendly applications, whereas the remaining 75% is stored or disposed into lagoons or landfills according to the regulations of CFPSs.

Since CFAs contains many toxic substances, metals, metalloids, and organic contaminates, its release to the atmosphere is hazardous to the biotic community. Hence, fly ash has to be stored properly, otherwise its degradation is the cause of many serious environmental issues and creates difficulties for native communities. The strict constraint of disposal, decrease in availability of space in landfills, and enhancing the disposal cost demands the necessity of fly ash for use in a green and economical way.

Hence, the ultimate valorisation of CFA into useful product materials in a sustainable way rather than disposal and storage is the ultimate viable solution to save our globe from drastic environmental contamination and open a new novel opportunity on an economic point of view. The viable utilization of CFA now receives serious attention to solve environmental issues in a sustainable way. In this process, the major challenge is detoxifying CFA and the extracted toxins have to be converted into useful products in safe conversion conditions [19, 20].

Without detoxification, the use of CFAs is the cause of harmful effects to our environment. Since CFAs are a good pozzolanic material, presently they can be effectively used as an important component for the manufacturing of cement. There are some useful applications of CFAs, which include grouting, sub-grade stabilization, asphalt filler, pavement base course, structural fill, general engineering fill, infill, and soil amendment. Now, huge research work is carried out by using fly ash to formulate adsorbent materials, both in aqueous and gaseous applications [21].

The modified CFA is now observed to be efficient for the removal of toxic metal ions from wastewater or gaseous pollutants. Now, extensive research work is going on for the treatment of contaminated water and in the near future it is going to be a viable promising alternative for water treatment technology. CFA is considered as an inexpensive source of aluminosilicate, where zeolites can be easily manufactured. These zeolites based on fly ash can be formed by using various processes such as a hydrothermal, sonication approach, or hydrothermal process assisted by the alkaline fusion, microwave irradiation, and multi-step treatment methods [22].

The formation of zeolite is normally dependent upon the reaction parameters, which includes pressure, temperature, the reagent concentration in the solution, pH, aging period, and process of activation along with the concertation of Al_2O_3 and SiO_2 present in the CFAs. The use of fly ash-based zeolites is the cause of the increase in revenue from coal fired power sectors with the reduction of cost related to the dumping of coal ashes.

Normally, the heterogeneous catalyst is observed to be more interesting due to its easy removal process to produce catalysts and facilitates better ways for the completion of the chemical reaction than homogeneous catalysts. The catalytic materials present in the heterogeneous catalyst depend upon active components and parameters and their interactions. The catalytic materials and supports are mainly SiO_2, MgO, Al_2O_3, and TiO_2. The CFA mainly contains Al_2O_3 and SiO_2, therefore exhibiting the required properties such as thermal stability for catalytic activity [23, 24].

9.3 Composition of Coal Fly Ash

CFAs contain various metal and metalloid oxides, which are in the order of $TiO_2 < K_2O < Na_2O < MgO < Fe_2O_3 < CaO < Al_2O_3 < SiO_2$ [25]. The chemical composition of fly ash obtained from various countries is represented in Table 9.1

On the basis of industrial application, coal origination, and chemical composition, CFAs are broadly divided into two classes: Class F and Class C (on the basis of the American Society for Testing and Materials (ASTM)). Class F is called a pozzolanic type, whereas Class C is called a cementitious type and both these possesses the properties of binding agents to manufacture clinker and slabs. Pozzolanic properties of the Class F type of fly ash are mainly due to the existence of Al_2O_3 and SiO_2. The average composition of a standard coal derived fly ash is given in Table 9.2.

Table 9.1 Chemical composition of fly ash produced in different countries [26].

Country name	Chemical composition of fly ash							
	SiO_2	CaO	Al_2O_3	Fe_2O_3	MgO	K_2O	Na_2O	TiO_2
Bulgaria	30.1–57.4	1.5–28.9	12.5-25.4	5.1-21.2	1.1-2.9	0.8-2.8	0.4-1.9	0.6–1
Australia	31.1–68.6	0.1–5.3	17–33	1–27.1	0–2	0.1–2.9	1.2–3.7	1.2–3.7
China	35.6–57.2	1.1–7	18.8–55	2.3–19.3	0.7–4.8	0.8-0.9	0.6–1.3	0.2-0.7
Canada	35.5–62.1	1.2–13.3	12.5–23.2	3–44.7	0.4–3.1	0.5–3.2	0.1–7.3	0.4–1.0
Europe	28.5-59.7	0.5-28.9	12.5-35.6	2.6-21.2	0.6-3.8	0.4-4.0	0.1–1.9	0.5–2.6
Germany	20–80	2–52	1-19	1–22	0.5–11	0-2	0-2	0.1–1
France	47–51	2.3-3.3	26–34	6.9-9.8	1.5–2.2	NA	2.3–6.4	NA
India	50.2–59.7	0.6-9.0	14–32.4	2.7–16.6	0.1–2.3	0.2–4.7	0.2–1.2	0.3–2.7
Japan	53.9–63	2.0-8.1	18.2–26.4	4.2-5.7	0.9–2.4	0.6-2.7	1.1–2.1	0.8-1.2
Italy	41.7–54	2.0-10	25.9–33.4	3–8.8	0-2.4	0-2.6	0–1	1-2.6
Korea	50-50.7	2.6–6.2	24.7-28.7	3.7–7.7	0.7-1.11	1.1	NA	NA
Russia	40.5–48.6	6.9-13.2	23.2–25.9	NA	2.6-4.0	1.9–2.6	1.2–1.5	0.5-0.6
Poland	32.2–53.3	1.2-29.9	4–32.2	4.5-8.9	1.2–5.9	0,2-3.3	0.2–1.5	0.6-2.2

(Continued)

Table 9.1 Chemical composition of fly ash produced in different countries [26]. (*Continued*)

Country name	Chemical composition of fly ash							
	SiO_2	CaO	Al_2O_3	Fe_2O_3	MgO	K_2O	Na_2O	TiO_2
Spain	41.5–58.6	0.3–11.8	17.6–45.4	2.6–16.2	0.3–3.2	0.2–4	0–1.1	0.5–1.8
Serbia	53.5–59.7	5.8–8.7	17.4–21	6–10.5	2–2.7	0.6-1.2	0.4-0.5	0.5–0.6
S. Africa	46.3–67	6.4–9.8	21.3–27	2.4–4.7	1.9–2.7	0.5–1	0–1.3	1.2-1.6
Turkey	37.9–57	0.2–27.9	20.5–24.3	4.1–10.6	1–3.2	0.4–3.5	0.1–0.6	0.6–1.5
USA	34.9–58.5	0.7–22.4	19.1–28.6	3.2–25.5	0.5–4.8	0.9–2.9	0.2–1.8	1-1.6
Maximum	80.0	52.0	55.0	44.7	11.00.0	4.7	7.3	3.7
Minimum	20.0	0.1	1.0	1.0	0.0	0.0	0.0	0.1

Table 9.2 Average composition of coal derived fly ash [27].

Components present in fly ash	Wt. %
SiO_2	68.43
Al_2O_3	14.78
CaO	3.80
K_2O	2.51
Fe_2O_3	7.91
SO_3	0.57
TiO_2	1.16
MgO	0.32
ZrO_2	0.1
P_2O_5	0.07
MnO	0.07
Rb_2O	0.05
SrO	0.05
ZnO	0.02
Y_2O_3	0.01

9.4 Economic Perspective of Biodiesel

In June 2005, the Department of US Energy published the cost of $0.89/l and $0.74/l for biodiesel and diesel, respectively, however the cost of both these are continuously enhanced with respect to time and maximum value is achieved in 2008 up to $1.22/l and $1.29/l for diesel and biodiesel, respectively. With the discovery of shale oil in 2009, the value of the diesel and biodiesel reduced to $0.58/l and $0.85/l, respectively.

Then, after that a further increase in cost is observed in range of $(1–1.1) ± 0.1/l and $(1.1–1.2) ± 0.1/l for diesel and biodiesel, respectively, and the increases in cost are still to be less up to date [28].

The cost of biodiesel production depends upon different factors as follows:

- The price of fossil oil and refining cost per liter in native region
- The annual cost of farming of the feedstock per unit hector area
- The fuel cost required for farming, processing, and final harvesting
- The fuel cost utilized as fertilizer on the agricultural land per unit hector area
- Labour cost of the whole process of production
- The kind of cultivated land and taxation of these lands
- The cost of catalysts, energy, and alcohol used during the process of production of biodiesel [29].

Among all the factors necessary for reduction of the cost of production of biodiesel, 80% of the total cost is associated with the cost of feedstock farming and the cost of processing varies with the kind of process used for extraction of biodiesel.

The cost of biodiesel production can be decreased by choosing a comparatively less cost of raw materials, including the oil extracted from the agricultural residues and waste of unused cooking oil rather than pure useable cooking vegetable oil. The use of diversified feedstock can promote the reduction of price and cause stable agricultural financial conditions. Again, the cost of processing of the cultivated feedstock mainly depends upon the kind of process used. The continuous or batch kind of catalyst and the process of integration are highly dependent upon the cost of processing [30].

One of the important processes of reducing the cost of production is the method of continuous transesterification with utilization of fixed bed reactors and the product obtained can be easily separated by using this process. The key factor of this technique is the higher capability of production and requirement of less reaction time. Now, it is found that the continuous process along with the addition of bifunctional catalysts is more effective and beneficial and stable than only using homogeneous catalysts.

The by-product of such kind of generated glycerol during the production of biodiesel is of about the range of \$0.04–0.06/l and the excess of CH_3OH obtained can be recovered, which can reduce the overall cost of this technique. Since biodiesel is eco-friendly and biodegradable, its demand is continuously increasing as a biofuel on the basis of environmental point of view. The production of biodiesel by using fly ash as a catalyst is cost effective and eco-friendly along with the reduction in the cost of production. The catalyst used for biodiesel production may be a heterogeneous or homogeneous catalyst, which influences the cost of production and usability [31].

The process of production of biodiesel is more viable and economical by using heterogeneous catalysts rather than a homogeneous catalyst along with the production of some useful by-products and product materials. The effectiveness of heterogeneity of the process depends upon several factors such as kinds of oil used, the molar ratio of alcohol to oil, type of catalyst or catalytic system used, temperature, and kinds of reactors used. The fly ash derived catalyst is hybrid in nature and facilitates the production of biodiesel. In spite of so many beneficial aspects of utilizing the heterogeneous catalysts for production of biodiesel, still there are some demerits of its utility, especially longer reaction time for completion of this process [32, 33].

9.5 Biodiesel from Fly Ash Derived Catalyst

9.5.1 Coal Fly Ash-Derived Sodalite as a Heterogeneous Catalyst

Biodiesel is presently produced by using a homogeneous base catalyst by triglycerides, which is the cause of synthesizing methyl esters of fatty acids. The homogeneous method catalysed by base is highly sensitive to the formation of fatty acids, which is the cause of less yield of biodiesel and the formation of soap. Therefore, in order to prevent saponification, the use of H_2SO_4 as a catalyst can catalyse simultaneously both transesterification and esterification reactions. Hence, biodiesel can be successfully produced from oil of low cost or waste oil containing higher fatty acids.

The acid catalyst used in the process of transesterification is found to be much slower compared to the basic catalyst and some extra methanol is needed for effective transesterification. In most of the transesterification reactions, heterogeneous catalyst is beneficial because it can be easily separated from the medium and again reused effectively in the subsequent process. There are a number of heterogeneous catalysts used in esterification and transesterification reactions such as Nb_2O_5, TiO_2 nanotubes, pre-treated silica (SiO_2HF), and $Sr_3Al_2O_6$ nanocomposite, etc. Zeolites are normally a crystalline substance of microporous aluminosilicates and highly effective in the process of both esterification and transesterification [34, 35].

Not only chemicals, but also some waste materials produced by many industries are the appropriate educts for the formation of zeolite. The fly ash produced due to burning of coal (coal fly ash) is one of the abundant waste materials from coal-based thermal power plants. Since this kind of fly ash contains mostly silica-aluminum, it can be easily and economically be converted into zeolite by the thermal treatment process in an alkaline solution. Such a treatment method is the cause of formation of different kinds of zeolites at various temperatures, alkaline reagent concentrations, reaction times, and ratios of the solution/fly ash [36].

Zeolites have proved as a promising catalyst in the process of transesterification of waste or refined oil. In the transesterification of soy oil by using ETS-10 zeolites and NaX, the rate of conversion achieved was 82.0% with the NaX at 60°C in 24h and it was studied that 95.1% methyl esters was obtained at a temperature of 60°C in 7h by using zeolite- X. However, heterogeneous transesterification cannot be achieved by the zeolite derived from coal fly ash, regardless of the reduction in reaction time, which is the vital path for achieving alkaline transesterification and feasibility of the process [37].

9.5.1.1 *Zeolite Synthesis from Coal Fly Ash*

The synthesis of zeolite is carried out by taking 0.15 g of NaOH with 16 ml of deionized water. This prepared solution is divided into two equal parts and kept in two air tight polypropylene bottles (Suppose solution A and B). To solution A, 1.65 g of $NaAlO_2$ was added and stirred continuously until fully dissolved. Then, in solution-B, 1.17 g of NaOH, 0.89 g of SiO_2 or coal fly ash and 1.3 mL of deionized water was added until fully dissolved. Then, the silicate solution B was added into aluminate solution B. Therefore, a sticky gel like substance was formed after stirring 15 minutes continuously because of full homogenization. The gel formed was kept up to the aging period of 6 days while maintaining a crystalline temperature of 100°C for 24h. Then, the mixture of the substance was clearly washed with deionized water until pH 9.0 was achieved [38].

9.5.1.2 *Production of Biodiesel through Heterogeneous Transesterification*

The process of transesterification reactions generally proceeds in a 100 ml flask fitted with a condenser. The required quantity of soybean oil was poured into the flask and then heated uniformly at a temperature of 65°C with vigorous stirring at a rate of 300 rpm. Then, catalyst and methanol both are added to the oil to initialize the reaction time. The total time for the required reaction maintained at about 2 hours. Then, when the reaction was complete, the reaction mixture was taken into a separating funnel where the biodiesel and glycerol phases are separated. Now, the glycerol phase with catalyst was again recovered and stored properly for the next use. After that, the biodiesel phase was finally purified by heating at a temperature of 55°C and stored [39].

9.5.2 CaO/Fly Ash Catalyst for Transesterification of Palm Oil in Production of Biodiesel

Palm oil is an important kind of vegetable oil and broadly available in nature in huge amounts, which is a good quality and low cost crop oil. In spite of its low cost, it can be used as a potential raw material for production of biodiesel. The major disadvantages of the palm oil are a higher percentage of free fatty acids (FFA) which can be reduced by using two stage reactions. The FFA released can be retarded via esterification reaction before proceeding to the transesterification reaction.

The FFA obtained permits for the transesterification reaction and can be used for the synthesis and use of heterogeneous catalysts for the production of biodiesel. The major advantages of utilizing the heterogeneous catalysts in transesterification reactions are eco-friendly towards the environment, non-corrosive, easy to separate, reusable, highly selective, and do not produce a huge amount of waste. CaO is considered a basic heterogeneous catalyst and can be developed in different forms such as CaO, $Ca\ (OH)_2$, and $CaCO_3$. The CaO/fly ash and CaO/Fe_3O_4 were used as an excellent catalyst for the production of biodiesel through the process of transesterification reaction [40, 41].

The application of CaO as a catalyst is not suitable because the O^{2-} ion formed due to its ionization on the surface can form hydrogen bonds with glycerol or methanol. This is the cause of enhancing the viscosity of the glycerol and forms suspension with CaO, which is very difficult for separation. Hence, to overcome these limitations, CaO can be reinforced by an active metal and the catalyst associated with CaO can be synthesized by using palm oil fly ash from the combustible products of light fractions of husk in the boiler. Fly ash can be utilized as a potential source of silica during the synthesis of catalyst, where it can function as an adsorbent at the active sites of the reaction. The presence of excessive concentrations of silica in CaO is the cause of enhancing the mechanical, morphology, and hardness properties, which leads to the prevention of leaching. Palm oil fly ash is a waste inorganic material having the chemical composition of

55.19% SiO_2, 2.12% Na_2O, 30.01% Al_2O_3, 1.91% MgO, 4.58% Fe_2O_3, 0.77% CaO, 1.40% K_2O, 2.74% TiO_2, and 1.28% BaO. Due to the presence of a higher percentage of Al_2O_3 and SiO_2, it can be treated as a potential low cost catalyst [42].

Generally, fly ash in association with CaO offers a huge surface area for the catalyst, which is the cause of better output of the catalyst. The catalyst was synthesized by the wet impregnation technology of fly ash and $CaCO_3$ in the composition of 55 wt.% and 45 wt.%, respectively. $CaCO_3$ was calcinated at a temperature of 800°C for about 1.5 hours. The catalyst was effectively applied in the process of transesterification of crude palm oil with CH_3OH in the process of the production of biodiesel. The ideal condition of transesterification is achieved at a temperature of 60°C in a stirring rate of 700 rpm, where the yield of biodiesel is about 75.73% [43].

CaO can be obtained due to the decomposition of $CaCO_3$ by the action of heat, but due to high temperature of calcination, more energy is needed for decomposition of $CaCO_3$, therefore another alternative is the fly ash and $Ca(NO_3)_2.4H_2O$. The major benefit of using $Ca(NO_3)_2.4H_2O$ is it can be directly used in higher temperatures without calcination and forms CaO. In the wet impregnation process, the composition of the catalytic system is 50 wt.% of fly ash and 50 wt.% $Ca(NO_3)_2.4H_2O$.

The ideal condition for the production of biodiesel was CaO/fly ash concentration of 70 wt.% of oil with a molar ratio of oil: methanol is 1:8 at a temperature of 60°C and the optimum yield biodiesel was 61.72% [44, 45].

9.5.2.1 Production of Biodiesel

The palm oil extracted is at first accurately weighed and then subjected into the esterification reactor. In that reactor, when the temperature reached 60°C, the reagent having oil had a methanol ratio of 1:12 and the required quantity of H_2SO_4 was added as a catalyst. Then, the reaction mixture was continuously stirred for about one hour at a speed of 400 rpm. The residual CH_3OH and mixture of H_2SO_4 catalyst are separated by using a separating funnel. The bottom layer is checked for FFA concentration and then the transesterification reaction starts. The bottom layer of the separated esterification product was then transferred into 50g of the transesterification reactor, then heated at a constant temperature of 70°C. After that, the reagent of oil with a methanol ratio of 6:1 and CaO/fly ash catalyst was added to it and stirred continuously at a speed of 400 rpm. Then, after 3 hours, the mixture is cooled, filtered, and the precipitate form of the catalyst is ultimately separated from the filtrate. Finally, the filtrate produced is subjected to purification and removal of impurity to get pure biodiesel [46].

9.5.2.2 *Transesterification Reaction*

The lowest % yield of biodiesel was achieved was 45.19% by using the catalytic system fly ash $Ca(NO_3)_2.4H_2O$ in the ratio of 1:9, where the concentration of CaO in the catalytic system is about 68.11%. The optimum % yield of biodiesel achieved was 71.77% by using the catalyst calcinated at a temperature of 800°C and the catalytic system fly ash $Ca(NO_3)_2.4H_2O$ is in the ratio of 2:8, where the concentration of CaO in the catalytic system is about 48.69%, which has the capability to provide a greater basic strength. The more the basic strength is, the higher the catalytic activity of the added catalyst and the higher the % yield of the biodiesel. At 800°C temperature of calcination, the added chemical structure of the catalyst does not change and is the cause of the deactivation of the catalyst. At a temperature of 850°C, the surface area of the catalyst is so large that all the active sites are unevenly distributed across the pore and do not react completely. On the other hand, the calcination at the temperature of 900°C with the catalytic system having the weight ratio of fly ash $Ca(NO_3)_2.4H_2O$ let down the catalytic activity and again with further increase in temperature of calcination, the active ends of the catalyst are deactivated and cause the agglomeration of catalyst. The excess of CaO over the required value in the catalyst has the possibility of retarding the decomposition of CH_3OH to CH_3O^- ion, therefore, the transesterification reaction cannot be initiated [47, 48].

9.5.3 Biodiesel Production Catalysed by Sulphated Fly-Ash

The combustion of coal in thermal power plants produces a huge amount of solid fly ash wastes, whose disposal is a problem and requires a huge land area and these ashes produced are the cause of significant water and air pollution. A major fraction of the fly ash is utilized for land recrimination, mines filling, and an important parameter for the manufacturing of cement, tiles, and bricks.

There are many reports available for development of catalyst by using fly ash, such as the framework of ash-metal-organics, ZSM-5 zeolite extracted from fly ash, sulphated zirconia loaded fly ash, etc. and their catalytic use in different organic transformations, valorisation of biomass products, and separation of nitrophenol from wastewater. It was found that the use of heterogeneous acid catalysts are considered the most viable and cost effective methods for the production of biodiesel from the feedstock of low price biomass containing free fatty acid (FFA) including industrial acid oil, animal fat, waste cooking oil, soap stock, and non-edible oil [49, 50].

Zeolite X, Zeolite-Y, Zeolite-Na X, Zeolite-A, Zeolite-K, sodalite, and fly ash based metal oxide catalysts are the most effective for production of biodiesel through the process of transesterification reaction. The use of zeolite of different types as catalysts, particularly the oil containing higher fatty acids feedstock, is utilized in the production of biodiesel and requires a high volume of water for ammonium exchange treatment of zeolite, which is the post-treatment process and is not noticeably effective [51].

The application of Titania and Zirconia as catalysts for the production of biodiesel is found to be highly effective, in which the Titania and sulphated Zirconia are found to be an excellent catalytic activity over their parent counterparts. Hence, it was observed that the sulphated solid acid catalyst extracted from fly ash is an excellent catalyst for production of biodiesel from feedstock of biowastes having high FFA and maize acid oil (MAO). The MAO is normally a by-product obtained during the refining of crude maize oil. During the refining process, a soap stock is found and that soap stock is converted into acid oil by adopting the process of acidulation. The different parameters that are responsible for optimization of the reaction include catalyst loading, the ratio of methanol to MAO, temperature, and time duration of the reaction [52, 53].

Table 9.3 Elemental analysis of SFA and FA [54].

Element present	% of SFA	% of FA
Oxygen (O)	41.58	34.66
Iron (Fe)	4.47	5.25
Manganese (Mn)	0.21	0.23
Magnesium (Mg)	0.35	0.30
Aluminium (Al)	13.3	21.86
Silicon (Si)	21.25	32.04
Sulphur (S)	15.77	0.0
Potassium (K)	0.94	1.80
Calcium (Ca)	0.62	1.35
Titanium (Ti)	1.52	1.92

The elemental analysis of sulphated fly ash (SFA) and fly ash (FA) is shown in Table 9.3.

9.5.4 Composite Catalyst of Palm Mill Fly Ash-Supported Calcium Oxide (Eggshell Powder)

Biodiesel generally consists of mono-alkyl esters, which are generally extracted from bio-resources of plants and animals containing long chain fatty acids which possess superior properties such as biodegradability, renewability, and nontoxicity. Normally, biodiesel was produced by the transesterification of high-grade vegetable oil by using homogeneous base catalysts effective for use. Some commonly identified important catalysts are NaOH, KOH and alkoxides, and carbonate salts.

These catalysts used are very difficult to separate and produce large volumes of waste water. The homogeneous catalyst is not suitable for production of biodiesel, therefore, scientists are searching the effectiveness of heterogeneous catalysts for the production of biodiesel. There are some heterogeneous catalysts such as SO_4^{2-}/TiO_2-SiO_2, $Sr(NO_3)_2/ZnO$, KI/Al_2O_3, $Na/NaOH/Al_2O_3$, KF/MgO, ion-exchange resins, etc. The use of CaO catalysts in reagent grade production of biodiesel is an expensive process, therefore scientists are searching for a low cost material. Chicken eggshell is a zero cost biomaterial and is normally found as the solid waste in egg shops and Indian restaurants. The chicken eggshell powder contains a high % of CaO. Not only chicken eggshell, but also oyster shells, mollusk shells, and mud crab shells are enriched with CaO and can be used as an alternative source of CaO in the production of biodiesel [55].

Palm mill fly ash is an important inorganic material obtained as the mill waste products of palm oil refineries having a chemical composition of 55.19% SiO_2, 2.74% TiO_2, 30.01% Al_2O_3, 2.12% Na_2O, 4.58% Fe_2O_3, 1.91% MgO, 0.77% CaO, 1.28% BaO, and 1.40% K_2O in the dry state. The excessively high percentage of Al_2O_3 and SiO_2 is the cause of catalytic support of palm mill fly ash for production of biodiesel. Presently, it was found that fly ash in association with CaO catalyst can become highly effective for Knoevenagel condensation along with good recyclability [56].

The palm oil fly ash saturated with $Ca(NO_3)_2.4H_2O$ was synthesized at a calcination temperature of about 800°C, the ratio of palm fly ash and $Ca(NO_3)_2.4H_2O$ is about 1:4, and oil/methanol ratio is 1:6 with reaction temperatures of 70°C with a reaction time is of around 2h. The prepared catalyst $Ca(NO_3)_2.4H_2O$/fly ash has a surface area of 24.342 m^2/g and the activity of the catalyst for the synthesis of biodiesel is almost 71.17%.

Hence, the catalytic system containing fly ash/chicken egg shell powder was found to be the most useful for production of biodiesel by using the process of transesterification of low grade palm oil. There are different parameters that are influencing the production process, such as time duration of calcination, temperature, and processes of dehydration [57].

9.5.4.1 Preparation of the CaO/PMFA Catalyst

At first, CaO in powdered form can be prepared from chicken egg shell by using a combination of the processes of calcination, dehydration and hydration. The dry mass product was properly washed with distilled water and after that, dried at a temperature of 105°C overnight through an oven. Then, the dried form of the biomass product was made into a homogeneous powdered form having a size of almost about 150μm by mechanical grinding. The powdered form of eggshell was then subjected to the calcination process at a temperature of from 850 to 900°C for about 2 to 3 hours. Then, a three-neck vessel was filled with 250ml of deionized water, followed by mixing of 18g of the calcinated powdered eggshells and then stirred continuously at a speed of 700 rpm. Then, 12g of palm mill fly ash (PMFA) was added in drop wise with continuous stirring for the next four hours to get the homogeneity of the solution. Now, the solution is kept for about 18 hours, where $Ca(OH)_2$ is precipitated on the PMFA. The presence of excess amounts of water was removed from the catalytic mixture, maintaining a temperature of approximately 105°C for about 24 hours. The dry form of catalyst was produced at this stage, which is then subjected again to recalcination at a temperature of 500-700°C in a muffle furnace for about 2-4 hours. Finally, the catalyst obtained is called as a CaO/PMFA catalysts and is highly useful in the process of transesterification in the preparation of biodiesel. By using this catalyst (CaO/PMFA catalyst), the maximum content of biodiesel produced is 86.2% of calcination temperature of 900°C for about 3 hours and at a dehydration temperature of 600°C for about 3 hours [58, 59].

9.5.5 Kaliophilite-Fly Ash Based Catalyst for Production of Biodiesel

Biodiesel is a sustainable alternative of renewable energy, which contains fatty acid alkyl esters and is an outstanding selection to replace fossil fuels. Presently, biodiesel is normally synthesized from oil of vegetable origin or animal origin by using various methods. Different kinds of homogeneous catalysis, enzyme catalysis, and heterogeneous catalysis are used for production of biodiesel. Among all these catalysis, a heterogeneous base catalysis system was found to be more popular because it is easy to separate and it has reusability in the subsequent steps.

The method of using conventional energy sources is a problem in our society because of the generation of a huge amounts of solid waste [60].

Example: Fly ash obtained due to combustion of pulverized coal. The combustion of low-grade fuels such as bituminous coal and coal gangue in thermal power plants, is commonly called a circulating fluidized bed fly ash (CFBFA) [61].

The untreated CFBFA is the cause of a great environmental problem. Hence, the CFBFA has to be recycled and modified to fit as a sustainable resource material. CFBFA can be converted into geopolymers which comprise of an amorphous to semi-crystalline three-dimensional inorganic network. Normally, geopolymers contain an agglomerate form of nanocrystalline zeolites and are compressed by an amorphous gel phase. The geopolymers and zeolites both are interconvertible to each other. In recent studies, it was found that geopolymers comprise of nanocrystals, which can be converted into zeolite of various structures by using hydrothermal reactions [62].

CFBFA-based geopolymers having amorphous structure are used to synthesize zeolite-like crystals or zeolite. Kaliophilite ($KAlSiO_4$) is usually a feldspathoid having a structure similar to zeolite containing the interlinking between $[AlO_4]^{5-}$ and $[SiO_4]^{4-}$ tetrahedra. Kaliophilite has hexagonal symmetry like nepheline. Now, kaliophilite is widely used for dehydrogenation of ethylbenzene to styrene, synthesis of NH_3, reforming of hydrocarbon vapour for production of H_2 and burning of diesel soot by using a catalyst because of its sites of more potassium active, which prevents the deposition of carbon. The presence of a number of basic potassium active sites, CH_3OH, and the property of insolubility in vegetable oil make kaliophilite a versatile heterogeneous catalyst for production of biodiesel [63, 64].

Kaliophilite was normally synthesized industrially by a solid phase reaction in between KOH and flint clay ($Al_2O_3{\cdot}2SiO_2{\cdot}2H_2O$) at a temperature of 1000°C. Presently, kaliophilite was synthesized by adopting the combined method of fusion, co-precipitation, and fast sol–gel technique. However, all these methods of heat treatment need temperatures in the range of 800–1200°C, along with requirements of high energy, which is the cause of the high cost of production. These problems can be overcome by converting geopolymer into kaliophilite, which has a composition similar to that of the zeolite and requires low-energy and low temperature in hydrothermal processes.

Kaliophilite is used as an excellent heterogeneous base catalyst for production of biodiesel, offering reutilization of CFBFA in an eco-friendly and sustainable manner on an industrial scale for the production of biodiesel in an economically viable method [65].

9.5.5.1 Synthesis of Kaliophilite

The usual method of the synthesis of kaliophilite normally includes the synthesis of CFBFA-based geopolymer and the hydrothermal conversion of geopolymer into kaliophilite. An alkali-activator having modulus of 1.5 was normally synthesized by dissolving KOH in the potassium water glass. After that ,100 g of CFBFA was mixed with 76.8 g of the activators and then stirred for about 5 min. Then, the whole mixture was taken in a 20 mm × 20 mm × 20 mm mold of steel packed in a bag made of polyethylene film. The geopolymer block of CFBFA base was found after curing at a temperature of 80°C for 24 hours. The hydrothermal process of transformation was generally carried out in an autoclave of 200 ml capacity and coated with liner (Teflon) containing a geopolymer monolith and 50ml of KOH solution at a temperature of 180°C for 24 hours. The monolithic kaliophilite having the compressive strength of 8 MPa was thoroughly washed, cleaned, and dried in an oven at a temperature of 105°C for 12 hours. Then, the block form of it was crushed and screened to get the granular form of kaliophilite catalyst, having particle sizes ranging from 0.15–0.315 mm [66, 67].

9.5.6 Fly-Ash Derived Zeolites for Production of Biodiesel

Zeolite is a significant mineral which contains SiO_4 and AlO_4 bonds. Zeolites are of various categories and possess three-dimensional crystalline structures, having the ability to permit diffusion and adsorption of molecules. A characteristic zeolite normally contains Al, O, and Si along with minerals of some specific metals such as Na, Mg, and K in suitable proportion having unique properties and structures.

Like other substances used as catalyst for production of biodiesel, zeolite is also highly effective, stable, and resistant to many adverse reaction conditions. Because of their high melting points (greater then 1000°C), they can withstand high reaction pressures and excellent stability in most of the solvents and air, if used as catalyst. The structural properties, acidity, and basicity of the zeolites are accountable for their advantage as well catalyst systems. Nowadays, zeolites can be designed and developed to a different degree of basicity or acidity, shapes to catalyze cracking, pore size, gasification, pyrolysis, and fuels production, and many kinds of reactions that are shaping selectively [68, 69].

Zeolites are obtained either naturally or can be prepared synthetically and the natural zeolites are widely found in different parts of our globe. It was estimated that more than 40 natural zeolite systems are found in sedimentary and volcanic rocks. Among these, the most commonly obtained zeolite minerals are chabazite, clinoptilolite, and mordenite. The natural zeolites are not fully fitted for industrial applications, therefore, on the basis of requirement, zeolites are prepared in the laboratory or on a commercial basis. More than 150 different kinds of zeolites are synthesized and later on successfully applied in various chemical syntheses including conversion of biomass.

For the synthesis of zeolites in an environmental friendly way, the use of fly ash is now becoming attractive because fly ash is a waste material produced in huge quantities due to huge industrial activity. The fly ash is a toxic solid waste material and is obtained as a waste product in different power plants because of the burning of coal. The siliceous and aluminous properties of fly ash are successfully treated as an excellent feedstock for the production of zeolite. It was estimated that every year huge amount of fly ash are produced from power plants and the quantity of production gradually increases very year because of an increase in the utilization of coal [70, 71].

In the year 2000, the production of fly ash was estimated to be more than 10 million tons and it drastically increased at a rapid rate in the subsequent years. About 50% of the generated fly ash was utilized in the manufacturing of cement, but the vital challenge towards the environment is inaccessibility of landfill for combustion and pollution issues. Hence, these problems are the cause of consideration in the synthesis of fly ash. The chabazite zeolites are synthesized by utilizing fly ash produced by the Japanese Coal Company. The three major factors in the synthesis of zeolites from fly ash by adopting an alkali hydrothermal process are:

- At first the fly ash obtained is dissolved in order to get Si^{4+} and Al^{4+} in the solution
- In the second step, aluminosilicate gel is formed via the condensation of ionic species obtained from the first step
- At last, crystallization occurred from the obtained gel to get the required property of zeolite crystals [72].

Since alkalis solutions like NaOH, Na_2CO_3, and KOH are used in this process, the ions CO_3^{2-} and OH^- released strongly promote the dissolution of the fly ash, otherwise the cations like K^+ and Na^+ significantly help or promote the process of crystallization from the gel structure of zeolite because of electrostatic interaction in between cations and anions. Zeolite A may be fashioned from fly ash through a one-step hydrothermal process. This method is advantageous because it neither required high temperature calcination nor long-term crystallization to produce crystals of zeolite A, having a crystallinity of 90.1-97.9% within 3 hours. Among the various zeolite structures, zeolite X in modification by using the alkali metals such as K, Cs, or Na can be used economically with suitable properties for synthesis of biodiesel [73]. There are many benefits of the synthesis of zeolite X such as:

- The cost of production of zeolite X from fly ash is about 20% less than the normal commercial production of zeolites
- The zeolite X is highly stable and can be modified and developed easily to gain the necessary activity. The K-X and Na-X zeolites are developed by using fly ash of Class F.
- The efficiency of the catalyst was measured by the trans esterification of sunflower oil at 65⁰C for 8 hours by the addition of catalyst of 3.0 wt. % and the maximum yield of biodiesel is of about 83.53%
- The activity of the catalyst is improved to increase in basicity of the zeolite and both of the catalysts were again reused for three cycles and also maintain outstanding performance [74]

Fly ash/NaOH in the ratio of 1:1 can be used for the synthesis of two kinds of zeolites, such as zeolites X and zeolites A, but the crystallinity of the composite material increases when the ratio increases to 1.5 [75].

Conclusion

The fly ash based catalyst was used in addition to 5% wt. KNO_3 or NaOH at a temperature of about 160°C, showing maximum conversion of oil (almost 86.13%). Again, the percentage conversion of oil was enhanced to increase in the catalyst loading. The maximum conversion of oil to biodiesel was achieved at a catalyst loading of 20%wt/wt. There are different process parameters, mainly reaction time, the molar ratio of methanol to oil, and reaction temperature, that greatly influence the production of biodiesel by using fly ash based material as a catalyst. Although the conversion of oil is not significantly increased with the increase in reaction time, the reaction time period of 5-8 hours provides maximum conversion of oil. Again, at the temperature of 200^{0}C, the conversion is found to be maximum. The use of fly ash based material as a catalyst for the production of biodiesel from different biomass feedstock is a cost effective, environmentally beneficial, and efficient method for production of biodiesel. The synthesis of heterogeneous fly ash-based catalysts promoted with alkaline earth and alkaline metals such as hydroxides, oxides, and their salts exhibits more interesting results because of the huge availability of alkaline earth or alkaline metal enriched waste materials and their corresponding catalytic action in the methanolysis process of triacylglycerol oils. The coal fly ash can be significantly modified and utilized as a solid heterogeneous catalyst for production of biodiesel. The transesterification by using the catalyst for the production of biodiesel by utilizing various feedstock biomass materials is highly satisfactory and interesting. The fly-ash derived zeolites, kaliophilite, CaO/PMFA, and palm mill fly ash associated with egg shell powder, sulphated fly-ash, and fly ash/CaO are found to be more beneficial, sustainable, cost effective, and efficient catalytic systems. More research work is still needed for development of the catalytic system based on fly ash for production of biodiesel, so that in addition to biodiesel production, the pollution load on our environment to some extent also is reduced.

References

1. Abas, N., Kalair, A., & Khan, N. 2015. Review of fossil fuels and future energy technologies. *Futures*, 69, 31–49. doi:10.1016/j.futures.2015.03.003
2. Huang, D., Zhou, H., & Lin, L. 2012. Biodiesel: an Alternative to Conventional Fuel. *Energy Procedia*, 16, 1874–1885. doi:10.1016/j.egypro.2012.01.287
3. Stephen, J. L., & Periyasamy, B. 2018. Innovative developments in biofuels production from organic waste materials: A review. *Fuel*, 214, 623–633. doi:10.1016/j.fuel.2017.11.042
4. Ambat, I., Srivastava, V., & Sillanpää, M. 2018. Recent advancement in biodiesel production methodologies using various feedstocks: A review. *Renewable and Sustainable Energy Reviews*, 90, 356–369. doi:10.1016/j.rser.2018.03.069
5. Chipurici, P., Vlaicu, A., Calinescu, I., Vinatoru, M., Vasilescu, M., Daniela Ignat, N., & Mason, T. J. 2019. Ultrasonic, Hydrodynamic and Microwave Biodiesel Synthesis–A Comparative Study for Continuous Process. *Ultrasonics Sonochemistry*. 57, 38-47. doi:10.1016/j.ultsonch.2019.05.011

6. Jamil, F., Al-Haj, L., Al-Muhtaseb, A. H., Al-Hinai, M. A., Baawain, M., Rashid, U., & Ahmad, M. N. M. 2018. Current scenario of catalysts for biodiesel production: a critical review. *Reviews in Chemical Engineer*ing, 34(2), 267–297. doi:10.1515/revce-2016-0026
7. Yao, Z. T., Ji, X. S., Sarker, P. K., Tang, J. H., Ge, L. Q., Xia, M. S., & Xi, Y. Q. 2015. A comprehensive review on the applications of coal fly ash. *Earth-Science Reviews,* 141, 105–121. doi:10.1016/j.earscirev.2014.11.016
8. Dwivedi, A., and Jain, M. K. 2014. Fly ash – waste management and overview: A Review. *Recent Research in Science and Technology*, 6(1), 30-35
9. Zierold, K. M. and Odoh, C. 2020. A review on fly ash from coal-fired power plants: chemical composition, regulations, and health evidence. *Reviews on Environmental Health*, 35(4), 401–418. DOI: https://doi.org/10.1515/reveh-2019-0039
10. Ahmaruzzaman, M. 2010. A review on the utilization of fly ash. *Progress in Energy and Combustion Science*, 36(3), 327–363. doi:10.1016/j.pecs.2009.11.003
11. Gollakota, A. R. K., Volli, V., & Shu, C.-M. 2019. Progressive utilisation prospects of coal fly ash: A review. *Science of The Total Environment*. 672, 951-989 doi:10.1016/j.scitotenv.2019.03.337
12. Patel, H. 2020. Environmental valorisation of bagasse fly ash:a review. RSC Adv., 10, 31611-31621. https://doi.org/10.1039/D0RA06422J
13. Blissett, R. S., & Rowson, N. A. 2012. A review of the multi-component utilisation of coal fly ash. *Fuel*, 97, 1–23. doi:10.1016/j.fuel.2012.03.024
14. Ohenoja, K., Pesonen, J., Yliniemi, J., & Illikainen, M. 2020. Utilization of Fly Ashes from Fluidized Bed Combustion: A Review. *Sustainability*, 12(7), 2988-3013. doi:10.3390/su12072988
15. Fuller, A., Maier, J., Karampinis, E., Kalivodova, J., Grammelis, P., Kakaras, E., & Scheffknecht, G. 2018. Fly Ash Formation and Characteristics from (co-)Combustion of an Herbaceous Biomass and Greek Lignite (Low-Rank Coal) in a Pulverized Fuel Pilot-Scale Test Facility. *Energies,* 11(6), 1581-1608. doi:10.3390/en11061581
16. Boycheva, S., Zgureva, D., Lazarova, K., Babeva, T., Popov, C., Lazarova, H., & Popova, M. 2020. Progress in the Utilization of Coal Fly Ash by Conversion to Zeolites with Green Energy Applications. *Materials*, 13(9), 2014-2029. doi:10.3390/ma13092014
17. Fauzi, A., Nuruddin, M. F., Malkawi, A. B., & Abdullah, M. M. A. B. 2016. Study of Fly Ash Characterization as a Cementitious Material. *Procedia Engineering*, 148, 487–493. doi:10.1016/j.proeng.2016.06.535
18. Vassilev, S. V., & Vassileva, C. G. 2007. A new approach for the classification of coal fly ashes based on their origin, composition, properties, and behaviour. *Fuel*, 86(10-11), 1490–1512. doi:10.1016/j.fuel.2006.11.020
19. Malik, A., & Thapliyal, A. 2009. Eco-friendly Fly Ash Utilization: Potential for Land Application. *Critical Reviews in Environmental Science and Technology*, 39(4), 333–366. doi:10.1080/10643380701413690

20. Brännvall, E.,& Kumpiene, J. 2016. Fly ash in landfill top covers–A review. *Environmental Science: Processes & Impacts*, 18(1), 11–21.
21. Koshy, N., Dondrob, K., Hu, L., Wen, Q., &Meegoda J. N. 2019. Mechanical Properties of Geopolymers Synthesized from Fly Ash and Red Mud under Ambient Conditions. *Crystals*, 9(11), 572- 585. https://doi.org/10.3390/cryst9110572
22. Govindasamy, B. R., Naik, V. K., & Balasundaram, A. (2020). Comparison of coronally advanced versus semilunar coronally repositioned flap in the management of maxillary gingival recessions. *The Saudi Dental Journal*. Article in press. doi:10.1016/j.sdentj.2020.05.005
23. Franus, W., Wdowin, M., & Franus, M. 2014. Synthesis and characterization of zeolites prepared from industrial fly ash. *Environmental Monitoring and Assessment*, 186(9), 5721–5729. doi:10.1007/s10661-014-3815-5
24. Dere Ozdemir, O., & Piskin, S. 2017. A Novel Synthesis Method of Zeolite X From Coal Fly Ash: Alkaline Fusion Followed by Ultrasonic-Assisted Synthesis Method. *Waste and Biomass Valorization*, 1-12. doi:10.1007/s12649-017-0050-7
25. Miricioiu, M. G., & Niculescu, V.-C. 2020. Fly Ash, from Recycling to Potential Raw Material for Mesoporous Silica Synthesis. *Nanomaterials*, 10(3), 474. doi:10.3390/nano10030474
26. Bhatt, A., Priyadarshini, S., Acharath Mohanakrishnan, A., Abri, A., Sattler, M., & Techapaphawit, S. (2019). Physical, chemical, and geotechnical properties of coal fly ash: A global review. *Case Studies in Construction Materials*, 11, e00263. doi:10.1016/j.cscm.2019.e00263
27. Manique, M. C., Lacerda, L. V., Alves, A. K., & Bergmann, C. P. 2017. Biodiesel production using coal fly ash-derived sodalite as a heterogeneous catalyst. *Fuel*, 190, 268–273. doi:10.1016/j.fuel.2016.11.016
28. Ogunkunle , O., Ahmed, N. A. 2019. A review of global current scenario of biodiesel adoption and combustion in vehicular diesel engines. *Energy Reports*, 5, 1560-1579. https://doi.org/10.1016/j.egyr.2019.10.028
29. Farid, M. A. A., Roslan, A. M., Hassan, M. A., Hasan, M. Y., Othman, M. R., & Shirai, Y. 2020. Net energy and techno-economic assessment of biodiesel production from waste cooking oil using a semi-industrial plant: A Malaysia perspective. *Sustainable Energy Technologies and Assessments*, 39,1-11, 100700. doi:10.1016/j.seta.2020.100700
30. Ma, Y., & Liu, Y. 2019. Biodiesel Production: Status and Perspectives. *Biofuels: Alternative Feedstock and Conversion Processes for the Production of Liquid and Gaseous Biofuels*, Elsevier, Chapter-21, 503–522. doi:10.1016/b978-0-12-816856-1.00021-x
31. Karmee, S., Patria, R., & Lin, C. 2015. Techno-Economic Evaluation of Biodiesel Production from Waste Cooking Oil—A Case Study of Hong Kong. *International Journal of Molecular Sciences*, 16(3), 4362–4371. doi:10.3390/ijms16034362

32. Lee, A. F., Bennett, J. A., Manayil, J. C., & Wilson, K. 2014. Heterogeneous catalysis for sustainable biodiesel production via esterification and transesterification. *Chem. Soc. Rev.*, 43(22), 7887–7916. doi:10.1039/c4cs00189c
33. Semwal, S., Arora, A. K., Badoni, R. P., & Tuli, D. K. 2011. Biodiesel production using heterogeneous catalysts. *Bioresource Technology*, 102(3), 2151–2161. doi:10.1016/j.biortech.2010.10.080
34. Malonda Shabani, J., Babajide, O., Oyekola, O., & Petrik, L. 2019. Synthesis of Hydroxy Sodalite from Coal Fly Ash for Biodiesel Production from Waste-Derived Maggot Oil. *Catalysts*, 9(12), 1052-165. doi:10.3390/catal9121052
35. Changmai, B., Vanlalveni, C., Ingle, A. P., Bhagatd, R. & Rokhum, L. 2020. Widely used catalysts in biodiesel production: a review. : *RSC Adv.*, 10, 41625-41679. DOI: 10.1039/d0ra07931f
36. Toniolo, N., & Boccaccini, A. R. 2017. Fly ash-based geopolymers containing added silicate waste. A review. *Ceramics International*, 43(17), 14545–14551. doi:10.1016/j.ceramint.2017.07.221
37. Shu, Q., Yang, B., Yuan, H., Qing, S., & Zhu, G. 2007. Synthesis of biodiesel from soybean oil and methanol catalyzed by zeolite beta modified with La^{3+}. *Catalysis Communications*, 8(12), 2159–2165. doi:10.1016/j.catcom.2007.04.028
38. Fukasawa, T., Karisma, A. D., Shibata, D., Huang, A.-N., & Fukui, K. 2017. Synthesis of zeolite from coal fly ash by microwave hydrothermal treatment with pulverization process. *Advanced Powder Technology*, 28(3), 798–804. doi:10.1016/j.apt.2016.12.006
39. Colombo, K., Ender, L., & Barros, A. A. C. 2017. The study of biodiesel production using CaO as a heterogeneous catalytic reaction. *Egyptian Journal of Petroleum*, 26(2), 341–349. doi:10.1016/j.ejpe.2016.05.006
40. Witoon, T., Bumrungsalee, S., Vathavanichkul, P., Palitsakun, S., Saisriyoot, M., & Faungnawakij, K. 2014. Biodiesel production from transesterification of palm oil with methanol over CaO supported on bimodal meso-macroporous silica catalyst. *Bioresource Technology*, 156, 329–334. doi:10.1016/j.biortech.2014.01.076
41. Helwani, Z., Fatra, W., Saputra, E., & Maulana, R. 2018. Preparation of CaO/Fly ash as a catalyst inhibitor for transesterification process off palm oil in biodiesel production. *IOP Conf. Series: Materials Science and Engineering* 334-344, 012077 doi:10.1088/1757-899X/334/1/012077
42. Helwani, Z., Ramli, M., Saputra, E., Putra, Y. L., Simbolon, D. F., Othman, M. R., & Idroes, R. 2020. Composite Catalyst of Palm Mill Fly Ash-Supported Calcium Oxide Obtained from Eggshells for transesterification of Off-Grade Palm Oil. *Catalysts*, 10(7), 724. doi:10.3390/catal10070724
43. Khan, I. U., Yan, Z., & Chen, J. 2019. Optimization, Transesterification and Analytical Study of Rhus typhina Non-Edible Seed Oil as Biodiesel Production. *Energies*, 12(22), 4290-4310. doi:10.3390/en12224290
44. Volli, V., Purkait, M. K., & Shu, C.-M. 2019. Preparation and characterization of animal bone powder impregnated fly ash catalyst for

transesterification. *Science of The Total Environment,* 669, 314–321. doi:10.1016/j.scitotenv.2019.03.080

45. Biswas, S., Katiyar, R., Gurjar, B. R., & Pruthia V. 2017. Biofuels and Their Production through Different Catalytic Routes. *Chem. Biochem. Eng. Q.*, 31 (1), 47–62
46. Zahan, K., & Kano, M. 2018. Biodiesel Production from Palm Oil, Its By-Products, and Mill Effluent: A Review. *Energies*, 11(8), 2132. doi:10.3390/en11082132
47. Widayat, Satriadi, H., Syaiful, Khaibar, A., & Almakhi, M. M. 2017. Biodiesel production by using heterogeneous catalyst from fly ash and limestone. *2017 International Conference on Sustainable Energy Engineering and Application* (ICSEEA). 41-44. doi:10.1109/icseea.2017.8267685
48. Hadiyanto, H., Lestari, S.P., Abdullah, A. 2016. The development of fly ash-supported CaO derived from mollusk shell of Anadara granosa and Paphia undulata as heterogeneous CaO catalyst in biodiesel synthesis. *International Journal of Energy Environmental Engineering 7,* 297–305. https://doi.org/10.1007/s40095-016-0212-6
49. Rizwanul Fattah IM, Ong HC, Mahlia TMI, Mofijur M, Silitonga AS, Rahman SMA and Ahmad A. 2020. State of the Art of Catalysts for Biodiesel Production. *Front. Energy Res.* 8, 101-117. doi: 10.3389/fenrg.2020.00101
50. Istadi, I., Anggoro, D. D., Buchori, L., Rahmawati, D. A., & Intaningrum, D. 2015. Active Acid Catalyst of Sulphated Zinc Oxide for Transesterification of Soybean Oil with Methanol to Biodiesel. *Procedia Environmental Sciences,* 23, 385–393. doi:10.1016/j.proenv.2015.01.055
51. Volli, V., & Purkait, M. K. 2015. Selective preparation of zeolite X and A from flyash and its use as catalyst for biodiesel production. *Journal of Hazardous Materials*, 297, 101–111. doi:10.1016/j.jhazmat.2015.04.066
52. Carlucci, C., Degennaro, L., & Luisi, R. 2019. Titanium Dioxide as a Catalyst in Biodiesel Production. *Catalysts*, 9(1), 75. doi:10.3390/catal9010075
53. Thanh, L. T., Okitsu, K., Boi, L. V., & Maeda, Y. 2012. Catalytic Technologies for Biodiesel Fuel Production and Utilization of Glycerol: A Review. *Catalysts,* 2(1), 191–222. doi:10.3390/catal2010191
54. Lathiya, D. R., Bhatt, D. V., & Maheria, K. C. 2019. Sulfated Fly-Ash Catalyzed Biodiesel Production from Maize Acid Oil Feedstock: A Comparative Study of Taguchi and Box-Behnken Design. Chemistry Select, 4(14), 4392–4397. doi:10.1002/slct.201803916
55. Helwani, Z., Ramli, M., Saputra, E., Putra, Y. L., Simbolon, D. F., Othman, M. R., & Idroes, R. 2020. Composite Catalyst of Palm Mill Fly Ash-Supported Calcium Oxide Obtained from Eggshells for Transesterification of Off-Grade Palm Oil. *Catalysts*, 10(7), 724-739. doi:10.3390/catal10070724
56. Mohamma dhosseini, H., Abdul Awal, A S M & Haq Ehsan, A. 2015 Influence of palm oil fuel ash on fresh and mechanical properties of self-compacting concrete. *Sadhan.* 40 (6), 1989–1999.

57. Dahdah, E., Estephane, J., Haydar, R., Youssef, Y., El Khoury, B., Gennequin, C., Aouad, S. 2019. Biodiesel production from refined sunflower oil over Ca-Mg-Al catalysts: Effect of the composition and the thermal treatment. *Renewable Energy*,146, 1246-1248. doi:10.1016/j.renene.2019.06.171
58. Marinković, D. M., Stanković, M. V., Veličković, A. V., Avramović, J. M., Miladinović, M. R., Stamenković, O. O., Jovanović, D. M. 2016. Calcium oxide as a promising heterogeneous catalyst for biodiesel production: Current state and perspectives. *Renewable and Sustainable Energy Reviews*, 56, 1387–1408. doi:10.1016/j.rser.2015.12.007
59. Yoosuk, B., Udomsap, P., Puttasawat, B., & Krasae, P. 2010. Improving transesterification activity of CaO with hydration technique. Bioresource Technology, 101(10), 3784–3786. doi:10.1016/j.biortech.2009.12.114
60. He, P. Y., Zhang, Y. J., Chen, H., Han, Z. C., & Liu, L. C. 2019. Low-energy synthesis of kaliophilite catalyst from circulating fluidized bed fly ash for biodiesel production. *Fuel*, 257, 116041. doi:10.1016/j.fuel.2019.116041
61. Zahedi, M., & Rajabipour, F. 2019. Fluidized Bed Combustion (FBC) fly ash and its performance in concrete. *ACI Materials Journal*, *116*(4), 163-172.
62. Bhandari, R., Volli, V., & Purkait, M. K. 2015. Preparation and characterization of fly ash based mesoporous catalyst for transesterification of soybean oil. *Journal of Environmental Chemical Engineering*, 3(2), 906–914. doi:10.1016/j.jece.2015.04.008
63. Zhuang, X. Y., Chen, L., Komarneni, S., Zhou, C. H., Tong, D. S., Yang, H. M., Yu, W. H., & Wang, H. 2016. Fly ash-based geopolymer: Clean production, properties and applications. *Journal of Cleaner Production*, 125, 253-267. https://doi.org/10.1016/j.jclepro.2016.03.019
64. Zaini, I. N., García López, C., Pretz, T., Yang, W., & Jönsson, P. G. 2019. Characterization of pyrolysis products of high-ash excavated-waste and its char gasification reactivity and kinetics under a steam atmosphere. *Waste Management*, 97, 149–163. doi:10.1016/j.wasman.2019.08.001
65. Wen, G., Yan, Z., Smith, M., Zhang, P., & Wen, B. 2010. Kalsilite based heterogeneous catalyst for biodiesel production. *Fuel*, 89(8), 2163–2165. doi:10.1016/j.fuel.2010.02.016
66. Novembre, D., Gimeno, D., d' Alessandro, N., & Tonucci, L. 2018. Hydrothermal synthesis and characterization of kalsilite by using a kaolinitic rock (Sardinia, Italy) and its application in the production of biodiesel. *Mineralogical Magazine*, 82(4),1–36. doi:10.1180/minmag.2017.081.080
67. Novembre, D., Gimeno, D. 2017. The Solid-State Conversion of Kaolin to KAlSiO4 Minerals: The Effects of Time and Temperature. *Clays Clay Miner.* 65, 355–366 . https://doi.org/10.1346/CCMN.2017.064077
68. Galadima, A., & Muraza, O. 2020. Waste materials for production of biodiesel catalysts: Technological status and prospects. *Journal of Cleaner Production*, 121358. doi:10.1016/j.jclepro.2020.121358
69. Aniokete, T.C., Ozonoh, M., & Daramola M.O. 2019. Synthesis of Pure and High Surface Area Sodalite Catalyst from Waste Industrial Brine and Coal Fly

Ash for Conversion of Waste Cooking Oil (WCO) to Biodiesel. International Journal of Renewable Energy Research. 9(4), 1924-1937

70. Xiao, M., Hu, X., Gong, Y., Gao, D., Zhang, P., Liu, Q.,Wang, M. 2015. Solid transformation synthesis of zeolites from fly ash. *RSC Advances*, 5(122), 100743–100749. doi:10.1039/c5ra17856h
71. Ojha, K., Pradhan, N.C. & Samanta, A.N.2004. Zeolite from fly ash: synthesis and characterization. *Bull Mater Sci* 27, 555–564. https://doi.org/10.1007/BF02707285
72. Amoni, B. de C., Freitas, A. D. L. de, Loiola, A. R., Soares, J. B., & Soares, S. de A. 2019. A method for NaA zeolite synthesis from coal fly ash and its application in warm mix asphalt. *Road Materials and Pavement Design*, 20(2), 1–10. doi:10.1080/14680629.2019.1633766
73. Hong, J. L. X., Maneerung, T., Koh, S. N., Kawi, S., & Wang, C.-H. 2017. Conversion of Coal Fly Ash into Zeolite Materials: Synthesis and Characterizations, Process Design, and Its Cost-Benefit Analysis. *Industrial & Engineering Chemistry Research,* 56(40), 11565–11574. doi:10.1021/acs.iecr.7b02885
74. Ren, X., Qu, R., Liu, S., Zhao, H., Wu, W., Song, H., Zheng, C., Wu, X., Gao X., 2020. Synthesis of Zeolites from Coal Fly Ash for the Removal of Harmful Gaseous Pollutants: A Review. *Aerosol and Air Quality Research*, 20, 1127–1144.
75. Sangita, K., Prasad, B., & Udayabhanu, G. 2016. Synthesis of Zeolite from Waste Fly Ash by Using Different Methods. *Asian Journal of Chemistry,* 28(7), 1435–1439. doi:10.14233/ajchem.2016.19682

10

Emerging Biomaterials for Bone Joints Repairing in Knee Joint Arthroplasty: An Overview

Shankar Swarup Das

Department of Farm Machinery and Power Engineering, College of Agricultural Engineering and Post-Harvest Technology (Central Agricultural University), Ranipool, India

Abstract

Currently, the number of primary knee joint arthroplasty (KJA) cases has increased annually and there are a substantial number of patients possessing the bone defects of the tibial plateau. How to select appropriate biomaterials for repairing the tibial plateau bone defects and acquire satisfactory results from primary KJA has gradually turned into a burning research area. Effective repair results are the basis of the initial stability of the knee joint implants, which confirms that patients can obtain satisfactory results within the life expectancy of the prosthesis. At present, there are a variety of materials used to repair tibial plateau bone defects in primary KJA. In this review, we summarized advantages and disadvantages of different materials used in repairing tibial plateau bone defects and hope to provide certain references for selection of appropriate materials during treatment. Proper biomaterials used to repair bone defects of the tibia plateau include traditional materials (structural bone grafts, bone cement, segmental metal augments, and tantalum cones) and new biomaterials (tissue-engineered bone, autologous platelet-rich plasma, and autologous chondrocytes). Such materials possess their own characteristics, advantages, and limitations. Traditional repair biomaterials generally tend to have difficulties like bone resorption, secondary breakdown, stress shielding, and implant loosening. Newly developed biomaterials have delivered new opportunities for treating these diseases during early stages. However, the clinical security and long-term effectiveness of these biomaterials must be further tested. Apposite

Email: shankarswarup@gmail.com

Prakash Kumar Sarangi (ed.) Biorefinery Production of Fuels and Platform Chemicals, (233–252) © 2023 Scrivener Publishing LLC

biomaterials should be selected for repairing the bone defects as per the precise size and depth of the tibial plateau.

***Keywords*:** Knee joint arthroplasty, tibial plateau, bone defect, replacement, repairing biomaterial

10.1 Introduction

With continuous improvement in the socioeconomic standard, patients with knee disorders demand improved quality of life and the number of knee joint arthroplasty (KJA) performed has increased annually [1]. Based on the trend in artificial joint replacement surgery performed in North America over the past 14 years, Singh *et al.* [2] made a bold prediction that the number of artificial knee replacement surgeries in North America will increase by 56% to 401% between 2020 and 2040 compared to present, indicating that there will be a potentially huge demand for artificial knee replacement surgery. The goal of KJA is to restore function to the knee joint [3–5] and improve quality of life of patients. With the increase in the number of patients undergoing primary KJA, a large number of patients developed variable degrees of bone defects of the tibia plateau [6, 7]. Improper treatment will seriously affect the initial stability of the knee prosthesis, resulting in poor long-term outcomes after surgery [8]. Current research mostly focuses on usage of biomaterials for repairing of the bone defects found at tibia plateau in order to provide a superior foundation for primary KJA and obtain satisfactory primary stability [9, 10]. Along with the current research in this field, various traditional biomaterials and novel materials have been used for repairing, which have deep inspiration on the treatment of bone defects at the tibia plateau. This review is intended to summarize the root causes, diagnosis, and classify the tibia plateau bone defects and to apply different biomaterials in repairing such defects, which aims to provide a reference for selection of the biomaterials for repairing the bone defects at the tibia plateau in primary KJA.

10.2 Resources and Selecting Criteria

A rigorous search was made from many electronic databases, including Web of Science, Chinese National Knowledge Infrastructure Database, and PubMed for articles investigating application of biomaterials for repairing the bone defects at the tibia plateau after primary KJA, published from September 2019. The search terms were as follows: total knee arthroplasty, tibial, bone defect, and bone repair material. A total of 1856 articles were discovered in the electronic database and after removal of duplicates, 100 studies were finally included in the review.

10.3 Reasons for Bone Defects of Tibia Plateau

There are many different causes for tibia plateau bone defects. From the perspective of knee joint motion trajectory, the knee flexion and extension movements are considered as the rolling, spinning, and gliding of the distal femur upon the tibial plateau [11]. The anterior or the posterior areas of the distal femoral condyle are larger than that of the tibial plateau, which can disperse pressure during movement, while the contact area on the tibia plateau is small, so the wear of the tibia plateau is more severe than that of the distal femur and different degrees of bone defects may occur [7]. From the perspective of diseases, the main causes for bone defects are as follows: (1) osteoarthritis (OA) when degeneration occurs, especially in patients with varus and valgus deformities of the knee [12], degeneration and wear of the cartilage, bone beneath the damaged cartilage that is exposed, and loss of cartilage protection can lead to irreversible bone defects; (2) Rheumatoid Arthritis (RA) [13, 14] is a chronic autoimmune disease that affects the joints, which is mainly located in the synovial and cartilage tissues of the joint, can lead to local bone tissue destruction and absorption, and eventually cause bone defects; (3) traumatic injuries of the joint [15] can lead to fractures around the knee, especially cause severe tibial plateau fractures, and the high-energy injuries not only produce severe cartilage damage, but also cause disturbance in the normal anatomical relationship, accelerate joint degeneration, and the occurrence of post-traumatic arthritis, which is often accompanied by severe bone defects in the later stages; (4) pathological bone destruction from tuberculosis, bacterial infections, tumors [16] occurring in tibial plateau can cause bone destruction, resulting in the occurrence of different degrees of bone defects; (5) iatrogenic factors such as improper handling of surgical instruments during the osteotomy that may cause convexity and concavity of osteotomy surfaces of the tibia plateau, leading to bone defects; (6) genetic factors presented by Ding *et al.* [17] found that bone defects can be caused by genetic factors. Under the same loading, some patients may be more likely to develop bone defects.

10.4 Classification of Bone Defects of Medial Tibia Plateau

The goal of treatment of tibial plateau bone defects [22] is to obtain lower limb alignment and maintain the balance between flexion and extension gaps through restoring the flatness of the articular surface. Tibial plateau bone defects are classified into different types according to the size and depth of bone defects by different scholars and corresponding treatment can be applied according to the types of bone defects in order to achieve the above-mentioned treatment goals. According to the classification proposed by Stokley *et al.* [23], bone defects are classified into contained and uncontained defects based on whether the peripheral cortex is involved. Uncontained defects can be divided into oblique and vertical defects according to the appearance of the defects. At present, the most widely used classifications include Rand and Anderson Orthopaedic Research Institute (AORI). The Rand classification [24] classifies the bone defects based on their depth and area, including type I (minimal, defect involving an area of <50% with a depth <5mm), type II (moderate, defect involving an area of 50% -70%, 5-10mm depth), type III (extensive, defect involving an area of 70% -90% area, depth ≥10mm), and type IV (massive cavitary, defect involving an area of > 90%). Type IV is subdivided into type A (intact peripheral cortical rim) and type B (deficient peripheral cortical rim) based on the presence or absence of an intact peripheral rim. AORI classification was proposed by Engh and Parks in 1997 for grading bone defects in revision TKA based on defect degree of metaphyseal involvement [25]. Currently, AORI classification is also applied to evaluate bone defects in primary TKA. AORI is a simple and practical system which can guide appropriate selection of prosthesis. Femoral and tibial bone defects can be classified according to the AORI classification. Bone defects of the tibial plateau are classified into three types (T1, T2, and T3): Type 1 defects are contained defects, the peripheral cortical rim is intact, and the joint line is maintained in its normal position. Type 2 defects are noncontained defects and the patellar ligament insertion or collateral ligament of the knee is intact.

Type 2 bone defects are subdivided into cortical bone defects of medial or lateral tibial plateau (type 2A) and cortical bone defects of both medial and lateral tibial plateaus (type 2B). Type 3 defects are severe metaphyseal bone defects involving the patellar ligament insertion which are often complicated with collateral ligament and patellar ligament injuries of the knee. AORI type 1 defects can be treated by increasing osteotomy depth to eliminate the bone defects and offsetting the prosthesis position to avoid the bone defect area and type 2 defects can be treated with repair materials to restore bone volume. Bone repair materials combined with knee arthroplasty using constrained condylar prosthesis are recommended for treatment of type 3 defects [26]. In addition to the common bone defect classification mentioned above, Bargar and Gross *et al.* proposed segmental, lacunar, central, and discontinuity defects according to the morphology of bone defects. Engh and Ammeen *et al.* [27] proposed a classification based on their own experience, which has not been widely used in clinical settings due to certain limitations.

10.5 Different Biomaterials for Tibial Plateau Bone Defects (Figure 10.1, Table 10.1)

Structural bone grafts can be classified according to their origin into autologous and allogeneic bone grafts. According to the appearance of the material, grafts can also be divided into block and particulate forms. Particulate autologous bone grafts can be used to repair bone cysts and small contained bone defects, especially defects with less than 5 mm of depth [28, 29]. Hanna *et al.* [30–32] reported that autogenous bone grafts have high tissue compatibility and no risk of immune rejection. Autogenous bone grafts use the patient's own bone harvested during osteotomy and are easy to obtain [33] at no additional cost. During the bone grafting procedure, impaction bone grafting [13, 31] can be used to compact the bone graft tightly, reduce the risk of subsequent collapse, and obtain good initial stability. For non-contained bone defects, previous studies [6, 34, 35] used particulate autologous bone grafts in combination with titanium mesh to repair the defects. Sugita *et al.* [29] used bone blocks harvested from the distal femur during osteotomy to reconstruct peripheral cortical defects, and the residual gaps were filled with particulate cancellous bone. The above mentioned bone graft procedures can increase the initial stability during repair of uncontained defects and good stability is beneficial for the bone grafts to promote bone tissue growth at the early stage. Once the bone grafts integrate with surrounding bone minerals, permanent repair can be achieved [36, 37]. However, the amount of bone harvested from a patient during surgery is limited. For repair of large bone defects, allogeneic bone grafts can be used [38]. There are allogeneic bone plates designed for large bone defects in the market, which can be used as structural grafts to support the prosthesis [39–41]. The allogeneic bone plate can be trimmed

Table 10.1 Traditional biomaterials used for repairing bone defects.

Biomaterials	Sources	Application method	Classification of bone defects	References
Structured bone grafts	Autologous or allogeneic bone	Particulate autologous bone grafts can be used to repair bone directly or used in combination with titanium mesh.	Rand Type I, AORI Type 1	[13, 28, 29, 31]
		Allogeneic bone block grafts can be used to fill bone defects directly or used in combination with screws or Kirschner wires.	Rand Type II, AORI Type 2	[39, 42]
Bone cement	Polymethyle methacrylate (PMMA)	Bone cement can be used alone to increase bone volume directly.	Rand Type-I, AORI Type 1	[18, 27, 56]
		Bone cement can also be used in combination with screws to provide good support.	Rand Type-II, AORI Type 2	[57, 60]

(*Continued*)

Table 10.1 Traditional biomaterials used for repairing bone defects. (*Continued*)

Biomaterials	Sources	Application method	Classification of bone defects	References
Segmental metal augments	Tantalum	Segmental metal augments can be freely assembled and used in combination with prosthesis with extension stems	Rand Type-II-III, AORI Type 2, 3	[63, 68, 73]
Tantalum cones	Tantalum	Tantalum cones should be used in combination with prosthesis with extension stems.	Rand Type III, AORI Type 3	[83, 84, 86]

according to the specific shape and size of the bone defects so that its shape can conform to the shape of different bone defects, which is particularly suitable for the treatment of large bone defects. However, large block bone grafts may displace and slide. Watanabe *et al.* [42] recommended the use of screws or Kirschner wires with bone grafts to improve the fixation, strengthen and increase the initial stability of the bone grafts, and prevent early loosening. Awadalla *et al.* [43–45] preformed a biomechanical study and showed that the combined use of tibial prosthesis with extension stem and bone grafts in the treatment of bone defects, especially defects of ≥10 mm depth [46], can share the pressure in the proximal tibial medullary cavity through the medullary cavity of the distal tibia and increase the initial stability of the bone graft. Dorr *et al.* [6] proposed the following methods to increase the initial stability of the bone graft: (1) the cartilage and fibrous tissues on the surface of the bone graft bed and bone graft material should be cleaned completely [47] and holes are drilled in the surface to facilitate local vascular reconstruction; (2) the prosthesis should completely cover the surface of the bone graft to reduce the occurrence of stress shielding and bone resorption; (3) for large bone defects, the cortical bone at the edges of the bone graft should be preserved and placed at lateral tibial plateau to share the longitudinal pressure; (4) the irregular bone defect is recommended to be trimmed to a step-like bone defect [47, 48] to reduce the sliding of the bone graft; (5) an operation should be performed carefully to reduce leakage of bone cement into spaces between bone graft materials. However, allogeneic bone grafts have certain disadvantages such as the risk of immune rejection, iatrogenic infection, and fusion failure [15, 49, 50].

Garino *et al.* [51] used autologous bone combined with allogeneic bone grafts to repair bone defects in order to eliminate their respective disadvantages, i.e., reducing the incidence of immune rejection by using autologous bone graft, and obtaining sufficient amount of bone by using allogeneic bone grafts. Bone cement can be used either alone or in combination with screws. Because the artificial knee prosthesis is fixed in the bone bed with bone cement, the use of bone cement for repair of tibial plateau defects does not result in any additional costs and it does not have the disadvantages of allogeneic bone grafts. Bone cement is suitable for bone defects of less than 10mm in depth [6], but if bone cement is used for bone

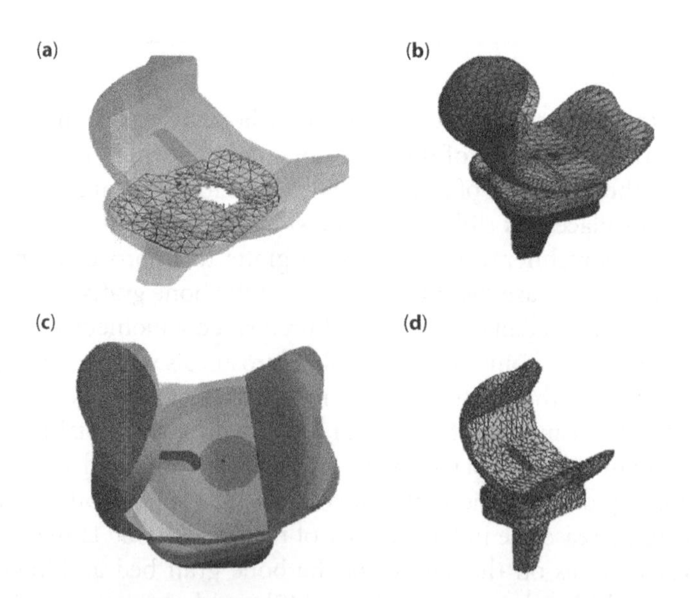

Figure 10.1 Bone Defects in Tibial Plateau and Treatment Methods. (a) During surgery, non-contained bone defects occurred in one side of the tibial plateau; (b) Bone grafts fixed with screws were used to repair non-contained bone defects; (c) Metal augments and tibial extension stems were used to repair non-contained bone defects; (d) Bone cement combined with rotating hinge knee prostheses were used to repair contained bone defects.

defects of >10 mm in depth, a tibial prosthesis can act on the cancellous bone area of the metaphysis directly [52–54], which may cause the occurrence of prosthesis loosening during weight-bearing exercises. Prosthesis loosening will cause an overall imbalance of the soft tissue surrounding the knee and the soft tissue balance plays an important role in postoperative stability [20, 55]. For bone defects smaller than 5mm, the articular surface can be trimmed during tibia osteotomy and covered with bone cement directly [18, 28, 34, 56]. When the depth of the defect is between 5-10mm, bone cement can be used in combination with screws to repair the defects. Freeman *et al.* [57] first reported the use of bone cement combined with screws for repairing bone defects in 1982. Berend *et al.* [58, 59] found that combined use of screws and bone cement in the repair of bone defects can increase local strength and screws can provide good support [57, 60] and simultaneously disperse stress at the contact area between artificial joints and bone cement. However, the screw has high stiffness and risk of breakage, which may fail to provide support. Therefore, Bilgen *et al.* [58, 61] used Kirschner wire to replace screws and obtained satisfactory results. Replacement of screws with Kirschner wire can effectively reduce risk of screw breakage and reduce the cost of consumables. However, bone cement has poor compatibility with surrounding bone tissue, low elastic modulus, and poor ability to withstand large loads, which may cause aseptic loosening of the prosthesis at the bone-to-cement interface in later stages after large longitudinal shear stress is applied [18, 19]. Prosthesis loosening accelerates the local wear of the prosthesis and affects the life span of prosthesis and revision surgery has to be performed [20, 21]. Therefore, the above methods are not recommended for treatment of large bone defects such as Rand type III and AORI type 3 defects [62].

Segmental metal augments are commonly used to fill large bone defects, especially AORI types 2 and 3 defects [63, 64]. Metal augments are suitable for uncontained defects and defects with a depth of more than 10 mm, but it is not recommended to treat defects exceeding 25 mm [65]. Metal augments can provide structural support and increased stability [66, 67] and can be flexibly combined with tibial prosthesis comprising intramedullary extension stems according to the size and depth of the defect [68]. Metal augments can be divided into half wedge-, wedge-, and ring-shaped augments [56], which can be freely combined according to different types of unilateral or bilateral bone defects in the tibial plateau. Baek *et al.* [69, 70] managed large bone defects by using double metal blocks on the medial aspect of the tibial component and obtained satisfactory results. The advantages of metal augments are that [71, 72] they are segmental and can be freely assembled, which can effectively reduce the difficulty of bone repair and improve surgical efficiency, the ability of metal augments to withstand pressure is better than bone graft materials and bone cement, use of metal augments can reduce the occurrence of height loss in the later stage after surgery, and they are able to avoid the risks of fusion failure and immune rejection. According to the mechanical study of the combined use of metal augments and tibial component intramedullary stems for bone defects, Whittaker *et al.* [63, 73, 74] found that during activity, pressure can be transmitted to the distal end of the knee prosthesis and the tibial intramedullary stems can share the pressure and effectively reduce prosthesis wear caused by activity. However, the metal augments do not produce osteogenic effects and stress shielding is likely to occur [73], resulting in disuse and osteoporosis of the surrounding bone. In addition, in order to match the shape of the metal augments, the surrounding bone should be trimmed, which can cause loss of bone quantity [75] and bring hidden dangers to the long-term stability of the prosthesis. Furthermore, if metal augments are larger than the bone defect, they will be exposed outside the cortex of the tibial plateau which can easily cut the collateral ligament, leading to injury and tear to collateral ligaments. Chung *et al.* [70] recommended the use of small metal augments to avoid cutting the ligaments around the knee joint, but small metal augments can reduce overall stability [55].

Tantalum cones were originally designed to reconstruct severe metaphyseal bone defects encountered during knee revision surgery [76, 77], but with the continued emergence of high-energy injuries, patients who have peri-articular fractures of the knee, especially severe tibial plateau fractures, are often complicated with severe bone defects. The traditional plate's internal fixation cannot allow patients to bear weight early and is often associated with severe traumatic arthritis [15]. In order to enable early weight-bearing, reduce the occurrence of complications (such as hypostatic pneumonia, bedsores, and deep venous thrombosis) caused by bed rest and immobilization, and avoid irreversible damage to normal knee tissue structures caused by traumatic arthritis [78–80], a considerable number of patients with knee fractures underwent one-stage artificial knee replacement surgery [81, 82]. Parratte *et al.* [83] used knee prosthesis with intramedullary stems in combination with tantalum cones to treat older patients with severe osteoporosis and found that the treatment cannot only repair bone defect of tibial plateau caused by high-energy injury [84, 85, 88], but also allow early weight-bearing activities and achieve satisfactory knee motion. Welldon *et al.* [86, 87] pointed out the following advantages of tantalum cones: (1) the elastic modulus of tantalum metal is closer to that of normal bone tissue of the human body and has excellent biocompatibility and safety; (2) tantalum with a cone shape design makes better contact with the surrounding bone, the inverted triangular step-like shape can disperse axial stress and partially share the load of the prosthesis; (3) tantalum cones can be fixed without cement, reducing the risk of loosening at the cement-to-bone interface in the later stage [84]; (4) segmental design of tantalum cones reduces surgical difficulty and saves operation time; (5) porous design of tantalum cones facilitates surrounding bone growth [89, 90] which can promote micro-interlocking of surrounding bone and prosthesis and increase the stability of the prosthesis; and (6) tantalum cones can be personalized using 3D printing according to the morphology of the prosthesis and bone defect, 3D-printed tantalum cones can match the prosthesis, avoid excessive grinding of the surrounding residual bone volume during surgery, and conform to the trend of precise and individualized treatment. Although the tantalum cones have the above advantages, their use in primary artificial knee replacement surgery is limited and can only be used to treat patients with massive bone defects (Rand type III and AORI type 3 defect) caused by severe trauma and tumors.

10.6 New Biomaterials to Repair Bone Defects in Tibia Plateau

With the emergence of the knee prosthesis, patients with knee pain have been effectively cured. However, due to the natural wear and tear of the prosthesis, many problems, such as dislocation and peripheral fractures may occur and prosthesis revision surgery may be required. It is actually a progressive process from knee cartilage destruction to bone defects. In order to delay the development of the disease, it is more effective to perform active intervention in the early stage of the disease than to perform surgery treatment after it has occurred. Ding *et al.* [93] showed that there was a correlation between tibial cartilage damage and bone defects and cartilage damage can be considered as a predictor of bone defects. A epidemiological study [94] found that the probability of knee OA is greatly increased in obese people, especially in women and their offspring and the causes may be that the small contact area of the knee increases the wear of cartilage, eventually leading to bone defects. With the deepening study of materials engineering, new types of repair materials are constantly emerging. (1) Autologous Chondrocytes: Vijayan *et al.* [95] showed that the use of autologous chondrocytes for patients with early knee OA, especially for young patients with cartilage defects, obtained good treatment efficacy and avoided the occurrence of bone defects and double-layer collagen membrane can be used in combination with autologous chondrocytes [96] to achieve sustained and effective release. However, the repair method is not effective in treating patients who had severe cartilage and bone defects, and those who had unhealthy habits, such as smoking and drinking [95]. (2) Autologous Platelet Rich Plasma (PRP): Autologous blood is separated by centrifugation to obtain PRP and then PRP is injected into the knee joint [97]. Autologous PRP can promote the repair of bone defects in the early stages of disease by enhancing cell proliferation and the synthesis of extracellular matrix. The source of PRP is highly safe. Use of PRP can avoid risks such as infection and immune rejection [98]. (3) Tissue-Engineered Bone: Human articular cartilage-derived sponge-like scaffolds are an oriented scaffold that fully mimic normal human tissues in terms of composition and tissue structure. Experimental studies have proved that chondrocytes can grow and differentiate on the scaffold in anticipated direction and can repair cartilage defects and mild bone defects and strength of the scaffold is superior to the above materials [99, 100]. Use of these new repair materials combined with minimally invasive technologies such as arthroscopy, microfracture, and drilling promote the regeneration of cartilage, delay the development of bone defects, to a certain extent, and have the advantage of relatively small iatrogenic injuries. However, these findings are largely based on results from animal experiments and there is a lack of large-scale clinical studies, so the clinical safety and long-term efficacy of the above materials are not clear.

10.7 Conclusion

There are still many controversies in regards to the treatment of tibial plateau bone defect in primary total knee joint arthroplasty, including the classification of bone defects, biomaterials, and application methods. Traditional materials, including structured bone grafts, bone cement, segmental metal augments, and tantalum cones possess their own characteristics, advantages, and shortcomings. There are no materials suitable for all types of bone defects. In general, for small tibial plateau bone defects, the treatment efficacy is better and the initial stability after surgery is high, while large bone defects are more difficult to repair and the stability of the prosthesis is relatively poor. Therefore, in order to improve knee joint alignment, stabilize the prosthesis, and reduce the occurrence of loosening, we should focus our attention on selection of appropriate repair materials according to the specific size and depth of the tibial plateau bone defect. During the treatment process, physicians should continuously summarize and develop personalized treatment plans for each patient. Furthermore, active intervention in the early stage of the disease, delaying primary KJA, and reducing the occurrence of bone defects are also current research directions. The emergence of new materials, including autologous chondrocytes, and tissue-engineered bone, has provided new opportunities to treat diseases during early stages. Through continuous summarization of traditional materials, continuous exploration of new materials and new treatment opinions, treatment of tibial plateau bone defects will no longer be a difficulty in primary KJA.

References

1. Kurtz SM, Ong KL, Lau E, Bozic KJ. Impact of the economic downturn on total joint replacement demand in the United States: updated projections to 2021. J Bone Joint Surg Am. 2014;96(8):624-30. doi: 10.2106/JBJS.M.00285.
2. Singh JA, Yu S, Chen L, Cleveland JD. Rates of Total Joint Replacement in the United States: Future Projections to 2020-2040 Using the National Inpatient Sample. J Rheumatol. 2019;46(9):1134-1140. doi: 10.3899/jrheum.170990.
3. Colizza WA, Insall JN, Scuderi GR. The posterior stabilized total knee prosthesis. Assessment of polyethylene damage and osteolysis after a ten-year-minimum follow-up. J Bone Joint Surg Am. 1995;77(11):1713-20.
4. Hawker G, Wright J, Coyte P, Paul J, Dittus R, Croxford R, Katz B, Bombardier C, Heck D, Freund D. Health-related quality of life after knee replacement. J Bone Joint Surg Am. 1998;80(2):163-73.
5. Victor J, Ghijselings S, Tajdar F, Van Damme G, Deprez P, Arnout N, Van Der Straeten C. Total knee arthroplasty at 15-17 years: does implant design affect outcome? Int Orthop. 2014;38(2):235-41. doi: 10.1007/s00264-013-2231-8.
6. Pei Z, Guan ZP, Zhang SL, Li YP, Zhang Z. [Autogeneous bone graft in the treatment of total knee arthroplasty for severe genu varus with tibial plateau bone defect]. Beijing Da Xue Xue Bao Yi Xue Ban. 2011;43(5):707-13.
7. Vahedian-Ardakani M, Mortazavi SM, Farzan M. Total Knee Arthroplasty: Does the Tibial Medial Side Defect Affect Outcome? Acta Med Iran. 2015;53(8):462-5.
8. Parratte S, Pagnano MW. Instability after total knee arthroplasty. J Bone Joint Surg Am. 2008;90(1):184-94.
9. Franceschina MJ, Swienckowski JJ. Correction of varus deformity with tibial flip autograft technique in total knee arthroplasty. J Arthroplasty. 1999;14(2):172-4.
10. Tsukada S, Wakui M, Matsueda M. Metal block augmentation for bone defects of the medial tibia during primary total knee arthroplasty. J Orthop Surg Res. 2013;8:36. doi: 10.1186/1749-799X-8-36.

11. Stiehl JB. A clinical overview patellofemoral joint and application to total knee arthroplasty. J Biomech. 2005;38(2):209-14.
12. Lu HS, Wang D. Reconstruction of proximal tibial oblique bone defect in total knee arthroplasty. Zhonghua Guke Zazhi. 2003;23 (8):446-469.
13. He Y, Ouyang GL, Xiao LB, Hu JL, Xia Q, Huang Z, Han DP, Zhu F, Sun ST. [Early efficacies of total knee arthroplasty for knees with severe lateral instability]. Zhonghua Yi Xue Za Zhi. 2013;93(43):3460-3
14. Cunnane G, FitzGerald O, Hummel KM, Youssef PP, Gay RE, Gay S, Bresnihan B. Synovial tissue protease gene expression and joint erosions in early rheumatoid arthritis. Arthritis Rheum. 2001; 44(8):1744-53.
15. Tigani D, Dallari D, Coppola C, Ben Ayad R, Sabbioni G, Fosco M. Total knee arthroplasty for post-traumatic proximal tibial bone defect: three cases report. Open Orthop J. 2011;5:143-50. doi: 10.2174/1874325001105010143.
16. Kassem Abdelaal AH, Yamamoto N, Hayashi K, Takeuchi A, Miwa S, Inatani H, Tsuchiya H. Ten-Year Follow-Up of Desarthrodesis of the Knee Joint 41 Years after Original Arthrodesis for a Bone Tumor. Case Rep Orthop. 2015;2015:308127. doi: 10.1155/2015/308127.
17. Ding C, Cicuttini F, Scott F, Stankovich J, Cooley H, Jones G. The genetic contribution and relevance of knee cartilage defects: case-control and sib-pair studies. J Rheumatol. 2005;32(10):1937-42.
18. Dorr LD, Ranawat CS, Sculco TA, McKaskill B, Orisek BS. Bone graft for tibial defects in total knee arthroplasty. Clin Orthop Relat Res. 1986;(205):153-65.
19. Faris PM, Ritter MA, Keating EM, Meding JB, Harty LD. The AGC all-polyethylene tibial component: a ten-year clinical evaluation. J Bone Joint Surg Am. 2003;85(3):489-93.
20. Ishii Y, Noguchi H, Matsuda Y, Kiga H, Takeda M, Toyabe S. Preoperative laxity in osteoarthritis patients undergoing total knee arthroplasty. Int Orthop. 2009;33(1):105-9.
21. Fehring TK, Odum S, Griffin WL, Mason JB, Nadaud M. Early failures in total knee arthroplasty. Clin Orthop Relat Res. 2001;(392):315-8.
22. Sculco TP. The role of constraint in total knee arthoplasty. J Arthroplasty. 2006;21(4 Suppl 1):54-6.
23. Stockley I, McAuley JP, Gross AE. Allograft reconstruction in total knee arthroplasty. J Bone Joint Surg Br. 1992;74(3):393-7.
24. Rand JA. Bone deficiency in total knee arthroplasty. Use of metal wedge augmentation. Clin Orthop Relat Res. 1991;(271):63-71.
25. Engh GA, Parks NL. The management of bone defects in revision total knee arthroplasty. Instr Course Lect. 1997;46:227-36.
26. Hu CC, Chen SY, Chen CC, Chang YH, Ueng SW, Shih HN. Superior Survivorship of Cementless vs Cemented Diaphyseal Fixed Segmental Rotating-Hinged Knee Megaprosthesis at 7 Years' Follow-Up.J Arthroplasty. 2017;32(6):1940-1945. doi: 10.1016/j.arth.2016.12.026

27. Engh GA, Ammeen DJ. Classification and preoperative radiographic evaluation: knee. Orthop Clin North Am. 1998;29(2):205-17.
28. Sohn JM, In Y, Jeon SH, Nho JY, Kim MS. Autologous Impaction Bone Grafting for Bone Defects of the Medial Tibia Plateau During Primary Total Knee Arthroplasty: Propensity Score Matched Analysis With a Minimum of 7-Year Follow-Up. J Arthroplasty. 2018; 33(8):2465-2470. doi: 10.1016/j.arth.2018.02.082.
29. Sugita T, Aizawa T, Miyatake N, Sasaki A, Kamimura M, Takahashi A. Preliminary results of managing large medial tibial defects in primary total knee arthroplasty: autogenous morcellised bone graft. Int Orthop. 2017;41(5):931-937. doi: 10.1007/s00264-016-3339-4.
30. Hanna SA, Aston WJ, de Roeck NJ, Gough-Palmer A, Powles DP. Cementless revision TKA with bone grafting of osseous defects restores bone stock with a low revision rate at 4 to 10 years. Clin Orthop Relat Res. 2011;469(11):3164-71. doi: 10.1007/s11999-011-1938-3.
31. Yoon JR, Seo IW, Shin YS. Use of autogenous onlay bone graft for uncontained tibial bone defects in primary total knee arthroplasty. BMC Musculoskelet Disord. 2017;18(1):502. doi: 10.1186/s12891-017-1826-4.
32. Toms AD, Barker RL, McClelland D, Chua L, Spencer-Jones R, Kuiper JH. Repair of defects and containment in revision total knee replacement: a comparative biomechanical analysis. J Bone Joint Surg Br. 2009;91(2):271-7. doi: 10.1302/0301-620X.91B2.21415.
33. Hosaka K, Saito S, Oyama T, Fujimaki H, Cho E, Ishigaki K, Tokuhashi Y. Union, Knee Alignment, and Clinical Outcomes of Patients Treated With Autologous Bone Grafting for Medial Tibial Defects in Primary Total Knee Arthroplasty. Orthopedics. 2017;40(4):e604-e608. doi: 10.3928/01477447-20170418-01.
34. Kharbanda Y, Sharma M. Autograft reconstructions for bone defects in primary total knee replacement in severe varus knees. Indian J Orthop. 2014;48(3):313-8. doi: 10.4103/0019-5413.132525.
35. Cai X, Wang Y, Wang JF, Zhou YG, Dong JY, Chen JY, Wei M, Wang ZG, Liu YJ, Li ZL. [Impacted cancellous autograft for reconstructing bone defects of tibial plateau in total knee arthroplasty]. Zhonghua Yi Xue Za Zhi. 2008;88(41):2907-11.
36. Ahmed I, Logan M, Alipour F, Dashti H, Hadden WA. Autogenous bone grafting of uncontained bony defects of tibia during total knee arthroplasty a 10-year follow up. J Arthroplasty. 2008;23(5):744-50. doi: 10.1016/j.arth.2007.08.021.
37. Sugita T, Aizawa T, Sasaki A, Miyatake N, Fujisawa H, Kamimura M. Autologous morselised bone grafting for medial tibial defects in total knee arthroplasty. J Orthop Surg (Hong Kong). 2015;23(2):185-9.

38. Bianchi G, Staals EL, Donati D, Mercuri M. The use of unicondylar osteoarticular allografts in reconstructions around the knee. Knee. 2009;16(1):1-5. doi: 10.1016/j.knee.2008.07.011.
39. Bush JL, Wilson JB, Vail TP. Management of bone loss in revision total knee arthroplasty. Clin Orthop Relat Res. 2006;452:186-92.
40. Morag G, Kulidjian A, Zalzal P, Shasha N, Gross AE, Backstein D. Total knee replacement in previous recipients of fresh osteochondral allograft transplants. J Bone Joint Surg Am. 2006;88(3):541-6.
41. Mounasamy V, Ma SY, Schoderbek RJ, Mihalko WM, Saleh KJ, Brown TE. Primary total knee arthroplasty with condylar allograft and MCL reconstruction for a comminuted medial condyle fracture in an arthritic knee--a case report. Knee. 2006;13(5):400-3.
42. Watanabe W, Sato K, Itoi E. Autologous bone grafting without screw fixation for tibial defects in total knee arthroplasty. J Orthop Sci. 2001;6(6):481-6.
43. Awadalla M, Al-Dirini RMA, O'Rourke D, Solomon LB, Heldreth M, Taylor M. Influence of varying stem and metaphyseal sleeve size on the primary stability of cementless revision tibial trays used to reconstruct AORI IIA defects. A simulation study. J Orthop Res. 2018;36(7):1876-1886. doi: 10.1002/jor.23851.
44. Toms AD, McClelland D, Chua L, de Waal Malefijt M, Verdonschot N, Spencer Jones R, Kuiper JH. Mechanical testing of impaction bone grafting in the tibia: initial stability and design of the stem. J Bone Joint Surg Br. 2005;87(5):656-63.
45. Awadalla M, Al-Dirini RMA, O'Rourke D, Solomon LB, Heldreth M, Rullkoetter P, Taylor M. Influence of stems and metaphyseal sleeve on primary stability of cementless revision tibial trays used to reconstruct AORI IIB defects. J Orthop Res. 2019;37(5):1033-1041. doi: 10.1002/jor.24232.
46. Mullaji AB, Padmanabhan V, Jindal G. Total knee arthroplasty for profound varus deformity: technique and radiological results in 173 knees with varus of more than 20 degrees. J Arthroplasty. 2005;20(5):550-61.
47. Donati D, Di Bella C, Lucarelli E, Dozza B, Frisoni T, Aldini NN, Giardino R. OP-1 application in bone allograft integration: preliminary results in sheep experimental surgery. Injury. 2008;39 Suppl 2:S65-72. doi: 10.1016/S0020-1383(08)70017-2.
48. Wilde AH, Schickendantz MS, Stulberg BN, Go RT. The incorporation of tibial allografts in total knee arthroplasty. J Bone Joint Surg Am. 1990;72(6):815-24
61. Bilgen MS, Eken G, Guney N. Short-term results of the management of severe bone defects in primary TKA with cement and K-wires. Acta Orthop Traumatol Turc. 2017;51(5):388-392. doi: 10.1016/j.aott.2017.02.002.
62. Lotke PA, Wong RY, Ecker ML. The use of methylmethacrylate in primary total knee replacements with large tibial defects. Clin Orthop Relat Res. 1991;(270):288-94.

63. Whittaker JP, Dharmarajan R, Toms AD. The management of bone loss in revision total knee replacement. J Bone Joint Surg Br. 2008;90(8):981-7. doi: 10.1302/0301-620X.90B8.19948.
64. Lee JK, Choi CH. Management of tibial bone defects with metal augmentation in primary total knee replacement: a minimum five-year review. J Bone Joint Surg Br. 2011;93(11):1493-6. doi: 10.1302/0301-620X.93B10.27136.
65. Pagnano MW, Trousdale RT, Rand JA. Tibial wedge augmentation for bone deficiency in total knee arthroplasty. A followup study. Clin Orthop Relat Res. 1995;(321):151-5.
66. Brooks PJ, Walker PS, Scott RD. Tibial component fixation in deficient tibial bone stock. Clin Orthop Relat Res. 1984;(184):302-8.
67. Hamai S, Miyahara H, Esaki Y, Hirata G, Terada K, Kobara N, Miyazaki K, Senju T, Iwamoto Y. Mid-term clinical results of primary total knee arthroplasty using metal block augmentation and stem extension in patients with rheumatoid arthritis. BMC Musculoskelet Disord. 2015;16:225. doi: 10.1186/s12891-015-0689-9.
68. Completo A, Sim?es JA, Fonseca F, Oliveira M. The influence of different tibial stem designs in load sharing and stability at the cement-bone interface in revision TKA. Knee. 2008;15(3):227-32. doi: 10.1016/j.knee.2008.01.008.
69. Baek SW, Choi CH. Management of severe tibial bony defects with double metal blocks in knee arthroplasty-a technical note involving 9 cases. Acta Orthop. 2011;82(1):116-8. doi: 10.3109/17453674.2010.548031.
70. Chung KS, Lee JK, Lee HJ, Choi CH. Double metal tibial blocks augmentation in total knee arthroplasty. Knee Surg Sports Traumatol Arthrosc. 2016;24(1):214-20. doi: 10.1007/s00167-014-3368-8.
71. Hube R, Pfitzner T, von Roth P, Mayr HO. [Defect Reconstruction in Total Knee Arthroplasty with wedges and blocks]. Oper Orthop Traumatol. 2015;27(1):6-16. doi: 10.1007/s00064-014-0331-2.
72. Qiu YY, Yan CH, Chiu KY, Ng FY. Review article: Treatments for bone loss in revision total knee arthroplasty. J Orthop Surg (Hong Kong). 2012;20(1):78-86.
73. Rawlinson JJ, Closkey RF Jr, Davis N, Wright TM, Windsor R. Stemmed implants improve stability in augmented constrained condylar knees. Clin Orthop Relat Res. 2008;466(11):2639-43. doi: 10.1007/s11999-008-0424-z.
74. Radnay CS, Scuderi GR. Management of bone loss: augments, cones, offset stems. Clin Orthop Relat Res. 2006;446:83-92.
75. Patel JV, Masonis JL, Guerin J, Bourne RB, Rorabeck CH.The fate of augments to treat type-2 bone defects in revision knee arthroplasty. J Bone Joint Surg Br. 2004;86(2):195-9.
76. Watters TS, Martin JR, Levy DL, Yang CC, Kim RH, Dennis DA. Porous-Coated Metaphyseal Sleeves for Severe Femoral and Tibial Bone Loss in Revision TKA. J Arthroplasty. 2017;32(11):3468-3473. doi: 10.1016/j.arth.2017.06.025.

77. Bugler KE, Maheshwari R, Ahmed I, Brenkel IJ, Walmsley PJ. Metaphyseal Sleeves for Revision Total Knee Arthroplasty: Good Short-Term Outcomes. J Arthroplasty. 2015;30(11):1990-4. doi: 10.1016/j.arth.2015.05.015.
78. Ye CY, Xue DT, Jiang S, He RX. Results of a Second-generation Constrained Condylar Prosthesis in Complex Primary and Revision Total Knee Arthroplasty: A Mean 5.5-Year Follow-up. Chin Med J (Engl). 2016;129(11):1334-9. doi: 10.4103/0366-6999.182845.
79. Weiss NG, Parvizi J, Hanssen AD, Trousdale RT, Lewallen DG. Total knee arthroplasty in post-traumatic arthrosis of the knee. J Arthroplasty. 2003;18(3 Suppl 1):23-6.
80. Bala A, Penrose CT, Seyler TM, Mather RC 3rd, Wellman SS, Bolognesi MP. Outcomes after Total Knee Arthroplasty for post-traumatic arthritis. Knee. 2015;22(6):630-9. doi: 10.1016/j.knee.2015.10.004. 81. Rai S, Liu X, Feng X, Rai B, Tamang N, Wang J, Ye S, Yang S. Primary total knee arthroplasty using constrained condylar knee design for severe deformity and stiffness of knee secondary to post-traumatic arthritis. J Orthop Surg Res. 2018;13(1):67. doi: 10.1186/s13018-018-0761-x.
82. Weiss NG, Parvizi J, Trousdale RT, Bryce RD, Lewallen DG. Total knee arthroplasty in patients with a prior fracture of the tibial plateau. J Bone Joint Surg Am. 2003;85(2):218-21.
83. Parratte S, Ollivier M, Argenson JN. Primary total knee arthroplasty for acute fracture around the knee. Orthop Traumatol Surg Res. 2018;104(1S):S71-S80. doi: 10.1016/j.otsr.2017.05.029.
84. Villanueva-Martínez M, De la Torre-Escudero B, Rojo-Manaute JM, Ríos-Luna A, Chana-Rodriguez F. Tantalum cones in revision total knee arthroplasty. A promising short-term result with 29 cones in 21 patients. J Arthroplasty. 2013;28(6):988-93. doi: 10.1016/j.arth.2012.09.003.
85. Troyer J, Levine BR. Proximal tibia reconstruction with a porous tantalum cone in a patient with Charcot arthropathy. Orthopedics. 2009;32(5):358.
86. Welldon KJ, Atkins GJ, Howie DW, Findlay DM. Primary human osteoblasts grow into porous tantalum and maintain an osteoblastic phenotype. J Biomed Mater Res A. 2008;84(3):691-701.
87. Nadorf J, Gantz S, Kohl K, Kretzer JP. Tibial revision knee arthroplasty: influence of segmental stems on implant fixation and bone flexibility in AORI Type T2a defects. Int J Artif Organs. 2016;39(10):534-540. doi: 10.5301/ijao.5000530.
88. Lizaur-Utrilla A, Collados-Maestre I, Miralles-Muñoz FA, Lopez-Prats FA. Total Knee Arthroplasty for Osteoarthritis Secondary to Fracture of the Tibial Plateau. A Prospective Matched Cohort Study. J Arthroplasty. 2015;30(8):1328-32. doi: 10.1016/j.arth.2015.02.032.
89. Huten D. Femorotibial bone loss during revision total knee arthroplasty. Orthop Traumatol Surg Res. 2013;99(1 Suppl):S22-33. doi: 10.1016/j.otsr.2012.11.009.

90. Levine BR, Sporer S, Poggie RA, Della Valle CJ, Jacobs JJ. Experimental and clinical performance of porous tantalum in orthopedic surgery. Biomaterials. 2006;27(27):4671-81.
91. Windhager R, Schreiner M, Staats K, Apprich S. Megaprostheses in the treatment of periprosthetic fractures of the knee joint: indication, technique, results and review of literature. Int Orthop. 2016;40(5):935-43. doi: 10.1007/s00264-015-2991-4.
92. Capanna R, Scoccianti G, Frenos F, Vilardi A, Beltrami G, Campanacci DA. What was the survival of megaprostheses in lower limb reconstructions after tumor resections? Clin Orthop Relat Res. 2015;473(3):820-30. doi: 10.1007/s11999-014-3736-1.
93. Ding C, Cicuttini F, Jones G. Tibial subchondral bone size and knee cartilage defects: relevance to knee osteoarthritis. Osteoarthritis Cartilage. 2007;15(5):479-86.
94. Ding C, Cicuttini F, Scott F, Cooley H, Jones G.Knee structural alteration and BMI: a cross-sectional study. Obes Res. 2005;13(2):350-61.
95. Vijayan S, Bentley G, Rahman J, Briggs TW, Skinner JA, Carrington RW. Revision cartilage cell transplantation for failed autologous chondrocyte transplantation in chronic osteochondral defects of the knee. Bone Joint J. 2014 ;96-B(1):54-8. doi: 10.1302/0301-620X.96B1.31979.
96. Vijayan S, Bartlett W, Bentley G, Carrington RW, Skinner JA, Pollock RC, Alorjani M, Briggs TW. Autologous chondrocyte implantation for osteochondral lesions in the knee using a bilayer collagen membrane and bone graft: a two- to eight-year follow-up study. J Bone Joint Surg Br. 2012;94(4):488-92. doi: 10.1302/0301-620X.94B4.27117.
97. Gobbi A, Lad D, Karnatzikos G. The effects of repeated intra-articular PRP injections on clinical outcomes of early osteoarthritis of the knee. Knee Surg Sports Traumatol Arthrosc. 2015;23(8):2170-2177. doi: 10.1007/s00167-014-2987-4.
98. Taniguchi Y, Yoshioka T, Kanamori A, Aoto K, Sugaya H, Yamazaki M. Intra-articular platelet-rich plasma (PRP) injections for treating knee pain associated with osteoarthritis of the knee in the Japanese population: a phase I and IIa clinical trial. Nagoya J Med Sci. 2018;80(1):39-51. doi: 10.18999/nagjms.80.1.39.
99. Adachi N, Ochi M, Deie M, Ishikawa M, Ito Y. Osteonecrosis of the knee treated with a tissue-engineered cartilage and bone implant. A case report. J Bone Joint Surg Am. 2007;89(12):2752-7.
100. Lin H, Zhou J, Cao L, Wang HR, Dong J, Chen ZR. Tissue-engineered cartilage constructed by a biotin-conjugated anti-CD44 avidin binding technique for the repairing of cartilage defects in the weight-bearing area of knee joints in pigs. Bone Joint Res. 2017;6(5):284-295. doi: 10.1302/2046-3758.65.BJR-2016-0277

About the Editor

Prakash Kumar Sarangi, PhD, is a scientist with a specialization in food microbiology at the Central Agricultural University in Imphal, Manipur, India. He has a PhD in microbial biotechnology from the Department of Botany at Ravenshaw University, Cuttack, India, an MTech degree in applied botany from the Indian Institute of Technology Kharagpur, India, and MSc degree in botany from Ravenshaw University, Cuttack, India. Dr. Sarangi's current research is focused on bioprocess engineering, renewable energy, biofuels, biochemicals, biomaterials, fermentation technology, and post-harvest engineering and technology. He has more than 10 years of teaching and research experience in biochemical engineering, microbial biotechnology, downstream processing, food microbiology, and molecular biology. He has served as a reviewer for many international journals and has authored more than 45 peer-reviewed research articles and 15 book chapters. Dr. Sarangi has edited five books by various publishers, including Wiley-Scrivener. Dr. Sarangi serves as an academic editor in the *PLOS One* journal. He is associated with many scientific societies as a Fellow Member (Society for Applied Biotechnology) and Life Member (Biotech Research Society of India, Society for Biotechnologists of India, Association of Microbiologists of India, Orissa Botanical Society, Medicinal and Aromatic Plants Association of India, Indian Science Congress Association, Forum of Scientists, Engineers & Technologists, International Association of Academicians and Researchers, Hong Kong Chemical, Biological & Environmental Engineering Society, International Association of Engineers, and Science and Engineering Institute).

Index

Also of Interest

Check out these other related titles from Scrivener Publishing:

BIOFUEL EXTRACTION TECHNIQUES, Edited by Lalit Prasad, Subhalaxmi Pradhan, and S.N. Naik, ISBN: 9781119829324. Written and edited by a team of world-class engineers, this groundbreaking new volume presents the current state of the art in the processing of feedstocks for biofuel production and end use application, extraction techniques, valorization of biofuel waste, and recent advances in biorefineries and the recycling industries.

"Advances in Biofeedstocks and Biofuels" series:

Volume 1:
BIOFEEDSTOCKS AND THEIR PROCESSING, Edited by Lalit Kumar Singh and Guarav Chaudhary, ISBN: 9781119117254. The most comprehensive and up-to-date treatment of all the possible aspects for biofeedstock processing and the production of energy from biofeedstocks.

Volume 2:
PRODUCTION TECHNOLOGIES FOR BIOFUELS, Edited by Lalit Kumar Singh and Gaurav Chaudhary, ISBN: 9781119117520. This second volume in the "Advances in Biofeedstocks and Biofuels" series focuses on the latest and most up-to-date technologies and processes involved in the production of biofuels.

Volume 3:
LIQUID BIOFUEL PRODUCTION, Edited by Lalit K. Singh and Gaurav Chaudhary, ISBN: 9781119459873. Focusing on renewable and carbon-cutting alternative fuel sources. this third volume in the "Advances in Biofeedstocks and Biofuels" series focuses on the latest and most up-to-date technologies and processes involved in the production of liquid biofuels.

Volume 4:
PRODUCTION TECHNOLOGIES FOR SOLID AND GASEOUS BIOFUELS, Edited by Lalit Kumar Singh and Gaurav Chaudhary, ISBN: 9781119785828. This latest volume in the series, "Advances in Biofeedstocks and Biofuels," offers the most up-to-date and comprehensive coverage available for the production technologies for solid and gaseous biofuels.

Other Related Titles

LIQUID BIOFUELS: Fundamentals, Characterization, and Applications, Edited by Krushna Prasad Shadangi, ISBN: 9781119791980. From the selection and pretreatment of raw materials to design of reactors, methods of conversion, selection of process parameters, optimization, and production of various types of biofuels to the industrial applications for the technology, this is the most up-to-date and comprehensive coverage of liquid biofuels for engineers and students.

BIODIESEL TECHNOLOGY AND APPLICATIONS, Edited by Inamuddin, Mohd Imran Ahamed, Rajender Boddula, and Mashallah Rezakazemi, ISBN: 9781119724643. This outstanding new volume provides a comprehensive overview on biodiesel technologies, covering a broad range of topics and practical applications, edited by one of the most well-respected and prolific engineers in the world and his team.

BIOFUEL CELLS: Materials and Challenges, Edited by Inamuddin, Mohd Imran Ahamed, Rajender Boddula, and Mashallah Rezakazemi, ISBN: 9781119724698. This book covers the most recent developments and offers a detailed overview of fundamentals, principles, mechanisms, properties, optimizing parameters, analytical characterization tools, various types of biofuel cells, edited by one of the most well-respected and prolific engineers in the world and his team.

Printed and bound by CPI Group (UK) Ltd, Croydon, CR0 4YY

12/06/2023

03226269-0001